I0054367

Composite Materials
SEAJCCM2024

South East Asia-Japan Conference on Composite Materials 2024, 13-15 August 2024, Kuala Lumpur, Malaysia

Editor
Norhayani Othman, Lin Feng Ng, Pui San Khoo, R.A. Ilyas, Mohd Yazid Yahya

Universiti Teknologi Malaysia, 81310 Johor Bahru, Johor, Malaysia

Peer review statement

All papers published in this volume of "Materials Research Proceedings" have been peer reviewed. The process of peer review was initiated and overseen by the above proceedings editors. All reviews were conducted by expert referees in accordance to Materials Research Forum LLC high standards.

Copyright © 2025 by authors

[(cc) BY] Content from this work may be used under the terms of the Creative Commons Attribution 3.0 license. Any further distribution of this work must maintain attribution to the author(s) and the title of the work, journal citation and DOI.

Published under License by **Materials Research Forum LLC**
Millersville, PA 17551, USA

Published as part of the proceedings series
Materials Research Proceedings
Volume 56 (2025)

ISSN 2474-3941 (Print)
ISSN 2474-395X (Online)

ISBN 978-1-64490-362-9 (Print)
ISBN 978-1-64490-363-6 (eBook)

This book contains information obtained from authentic and highly regarded sources. Reasonable efforts have been made to publish reliable data and information, but the author and publisher cannot assume responsibility for the validity of all materials or the consequences of their use. The authors and publishers have attempted to trace the copyright holders of all material reproduced in this publication and apologize to copyright holders if permission to publish in this form has not been obtained. If any copyright material has not been acknowledged please write and let us know so we may rectify in any future reprint.

Distributed worldwide by

Materials Research Forum LLC
105 Springdale Lane
Millersville, PA 17551
USA
https://mrforum.com

Manufactured in the United State of America
10 9 8 7 6 5 4 3 2 1

Table of Contents

Preface

Welcome to SEAJCCM 2024

The South East Asia-Japan Conference on Composite Materials 2024 (SEAJCCM 2024) with theme of "Composite Materials and Structures Toward Sustainable Future" was held with aims to provide a platform for knowledge sharing in research and recent technology of composites in various aspects. It is a relevant platform for academics, researchers, policymakers and private companies to collaborate and to discuss issues related to research related to composite science and technology where more than 40 organizations and 20 countries participated including the Bangladesh, United States of America, Italy, Iran, Ireland, Kingdom of Bahrain, Saudi Arabia, Czech Republic, Nigeria, United Kingdom, Ethiopia, Russia, Iraq, France, Pakistan, India, Indonesia, Singapore, Thailand and Malaysia. During this conference, the works about advanced manufacturing, applications, and recent developments on composite science and technology are presented. The presentations comprised different fields related to the composite science and technology including the natural fibres, mineral filler/recycling polymer composites, advanced metal matrix composites, fabric/film/rubber/polymer, and advanced polymer composites.

In the preparation of this conference, we have been helped by many colleagues and sponsors. We wish to acknowledge the supports given by our sponsors. In addition, we would like to thank all the SEAJCCM 2024 committee and volunteers, who have been sharing their time and strength in succeeding this conference from scratch. Last, but not least, we wish to thank the students, undergraduate, graduate, postdocs, researchers, and lecturers from various continents, who have joined our conference and shared their knowledge regarding their research activity with stimulating curiosity.

Organizing Committee

**South East Asia-Japan Conference on Composite Materials 2024
(SEAJCCM 2024)**

Advisor I	Prof. Ts. Dr. Mohd Hafiz Dzarfan bin Othman
Advisor II	Assoc. Prof. Dr. Shukur bin Hj. Abu Hassan
Chairman	Prof. Dr. Mohd Yazid bin Yahya
Co-Chairman	Prof. Dr. Mat Uzir bin Wahit
Secretary	Ts. Dr. Ahmad Ilyas bin Rushdan
Co-Secretary	Ms. Hafizah binti Ithnin
Treasurer	Ts. Dr. Muhammad Asyraf bin Muhammad Rizal
Co-Treasurer	Ms. Hafizah binti Ithnin
UNIT	
Registration	Assoc. Prof. Dr. Rohah binti A. Majid
	Assoc. Prof. Ir. Dr. Nadia binti Adrus
Publicity, Media, Promotion, Sponsorship, Awards & Certificate	Dr. Nur Hafizah binti Abd Khalid
	Assoc. Prof. Dr. Tuty Asma Abu Bakar
	Dr. Mohd Haziq bin Dzulkifli
	Dr. Muhammad Syahir bin Sarkawi
	Dr. Mohammad Fikry bin Mohammad Jelani
	Dr. Mohd Ayub bin Sulong
Technical & Logistic	Assoc. Prof. Dr. Mohamad Zaki bin Hassan
	Prof. Dr. Astuty binti Amrin
	Assoc. Prof. Ts. Dr. Roslina binti Mohammad
Publication & Proceeding	Dr. Norhayani Bt Othman
	Dr. Ng Lin Feng
	Dr. Khoo Pui San
Program, Protocol & Speakers	Dr. Muhamad Fauzi bin Abd Rased
	Dr. Nur Hafizah binti Ab Hamid
	Dr. Shuhada Atika binti Idrus Saidi
	Dr. Izni Mariah binti Ibrahim

Composite Materials: SEAJCCM2024   Materials Research Forum LLC
Materials Research Proceedings 56 (2025) 1-10   https://doi.org/10.21741/9781644903636-1

Innovative Hematite-Incorporated Geopolymer Membrane from Coal Fly Ash through Direct Foaming Method

Rendy Muhamad IQBAL[1,2,3,a] *, Retno AGNESTISIA[2,b],
Deni Shidqi KHAERUDINI[4,c], Elfrida Roulina SIMANJUNTAK[2,d],
Hamzah FANSURI[5,e], Muthia ELMA[6,7,f], Mohd Akmali MOKHTER[1,3,g] and
Mohd Hafiz Dzarfan OTHMAN[3,h]

[1]Department of Chemistry, Faculty of Science, Universiti Teknologi Malaysia, Skudai, Johor Bahru 81310, Malaysia

[2]Department of Chemistry, Faculty of Mathematic and Natural Science, Universitas Palangka Raya, Palangka Raya 73111, Indonesia

[3]Advanced Membrane Technology Research Center, Universiti Teknologi Malaysia, Skudai, Johor Bahru, 81310, Malaysia

[4]Research Center for Advanced Materials, National Research and Innovation Agency (BRIN), Kawasan Puspitek Serpong, South Tangerang, Banten 15314, Indonesia

[5]Department of Chemistry, Faculty of Science and Data Analytics, Institut Teknologi Sepuluh Nopember, Surabaya 60111, Indonesia

[6]Materials and Membrane Research Group, Lambung Mangkurat University, Banjarbaru 70714, Indonesia

[7]Department of Chemical Engineering, Faculty of Engineering, Lambung Mangkurat University, Banjarbaru 70714, Indonesia

[a]iqbal.rm@mipa.upr.ac.id, [b]retno.agnestisia@mipa.upr.ac.id, [c]deni.shidqi.khaerudini@brin.go.id, [d]simanjuntak.elf@gmail.com, [e]h.fansuri@chem.its.ac.id, [f]melma@ulm.ac.id, [g]mohdakmali@utm.my, [h]hafiz@petroleum.utm.my

Keywords: Fly Ash, Geopolymer, Hematite, Inorganic Membrane, Direct Foaming

Abstract. Fly ash generation is a continuously growing waste or by-product resulting from the combustion of coal, which is potentially to explore for membrane materials. The addition of hematite as an antifouling agent allows for the utilization of fly ash as a geopolymer-based membrane. The objective of this study is to create a hematite/geopolymer membrane utilizing type C fly ash by the direct foaming technique in order to enhance the porosity of the membrane. A homogeneous mixture was obtained by combining 65 grams of fly ash and 0.85 grams of Al(OH)$_3$ with a base activator. Subsequently, 3 grams of hematite were added to the mixture. Subsequently, hydrogen peroxide (H$_2$O$_2$) was added as a foaming agent and gradually blended with different percentages of 0, 2, 4, and 6 weight percent (wt.%) into the paste mixture. The paste mixture was put into the mould and left to undergo the process of curing for a duration of 7 days. The fly ash was analysed using X-ray fluorescence (XRF), X-ray diffraction (XRD), and particle size analysis. Next, the membrane composite underwent characterization using XRD, FTIR, scanning electron microscope (SEM), and Archimedes techniques. The findings indicated that the fly ash from Pulang Pisau's Power Plant was officially identified and categorized as type C class. The diffractogram revealed the presence of mullite, quartz, and hematite phases in the hematite/geopolymer membrane. This finding was further verified by the FTIR analysis, which detected molecular vibrations indicating the existence of the T-O-T (T= Si or Al) and O-H components of the geopolymer structure. The scanning electron microscope images reveal that the

Content from this work may be used under the terms of the Creative Commons Attribution 3.0 license. Any further distribution of this work must maintain attribution to the author(s) and the title of the work, journal citation and DOI. Published under license by Materials Research Forum LLC.

membrane surface exhibits a rough texture and contains a limited number of microcracks resulting from the direct foaming process. Additionally, the measurements of porosity indicate that the porosity of the membrane increases proportionally with increased concentrations of H_2O_2. This research demonstrates the potential of coal fly ash as a sustainable raw material for developing advanced membranes for wastewater treatment.

Introduction

Fly ash is a residue produced by the combustion of coal in a steam power plant. It contains a significant amount of chemical compounds, particularly SiO_2 and Al_2O_3, which can be utilized in the production of geopolymers [1]. The global production of fly ash is steadily increasing each year, with coal power plants continuing to be the primary source of electricity in many nations [2]. The larger quantity of this trash must be repurposed as value-added materials. Therefore, it is imperative to utilize fly ash for the goal of promoting sustainable development. One promising application of fly ash is its usage as a raw material in the production of inorganic or geopolymer membranes.

A novel inorganic membrane in development is a geopolymer-based membrane, which shows promise as an efficient material for water treatment. Geopolymers are inorganic polymers made up of aluminosilicate. They consist of a three-dimensional network of SiO^{4-} and AlO^{4-} tetrahedrons that share oxygen atoms as bridges [3]. Geopolymer materials exhibit favorable mechanical properties, possess excellent thermal stability, are readily manufacturable, cost-effective, and environmentally benign [4]. Geopolymers have diverse applications, including their usage as construction materials [5], adsorbents [3], and for immobilizing heavy metals [6], among others. Researchers have recently begun to explore the use of geopolymer as an inorganic membrane material. Faradilla et al. [7] have created a kaolin-based TiO_2/geopolymer membrane for the combined processes of separation and photocatalytic degradation of dyes. Subsequently, Chen et al. [8] fabricated a Cr_2O_3/geopolymer membrane with an identical purpose to that of Faradilla et al. [7]. Alternatively, Naveed et al. [9] created a geopolymer using fly ash, categorized as type F, for the purpose of water desalination. Subaer et al. [10] attempted saltwater desalination using a membrane made of TiO_2-rGO/geopolymer. They used a process called membrane pervaporation and achieved a salt rejection rate of up to 91%. The use of geopolymer membranes has great potential in the fields of wastewater treatment and water desalination.

A challenge frequently encountered while utilizing membranes based on inorganic materials for the filtration process is the tendency of fouling. The presence of high levels of organic pollutants or bacteria in polluted water leads to the attached on the surface and within the pores of the membrane [11]. Membrane fouling will affect the functionality and efficiency of the membrane. To prevent fouling on the membrane, it is necessary to incorporate a metal oxide, such as hematite (Fe_2O_3), into geopolymer-based membranes to enhance their performance [11,12]. Remarkably, hematite exhibited low toxicity, affordability, environmental friendliness, and greater stability in comparison to other forms of iron oxide [13,14]. In addition, it possesses hydrophilic surface qualities that effectively minimize membrane fouling. Moreover, it is cost-effective, exhibits great compatibility, and can be synthesized from readily available natural resources or industrial waste such as mill scale [15,16].

Furthermore, it is commonly observed that geopolymer-based materials have a low porosity [17,18]. The direct foaming approach shows promise in enhancing membrane porosity due to its simplicity, cost-effectiveness, and one-step synthesis process. As reported by Xia et al. [19], direct foaming approach was firstly introduced by Sundermann in 1973, which involves the use of inorganic or organic compounds as foaming agents to generate volatile gases and bubbles. The presence of gas and bubbles enhances the process of membrane pore creation, resulting in accelerated pore formation. Therefore, it has the capacity to enhance the efficiency of fly ash-based

Composite Materials: SEAJCCM2024
Materials Research Proceedings 56 (2025) 1-10

Materials Research Forum LLC
https://doi.org/10.21741/9781644903636-1

geopolymer membranes, as well as membrane composites, in certain applications such as wastewater treatment.

This study investigates the production of a composite material consisting of a hematite/geopolymer membrane. The membrane was fabricated using type c fly ash collected from the Pulang Pisau power plants in the Central Kalimantan Province of Indonesia. The fly ash was served as a source of alumina and silica, which are crucial components in the geopolymerization process. Hematite was added as an anti-fouling agent, and H_2O_2 was included to enhance membrane porosity in this study.

Methodology

Materials

The materials used in this work were fly ash collected in Pulang Pisau's power plant (Central Kalimantan Province, Indonesia), Na_2SiO_3 was supplied by Meteora Pelangi Jaya Company, $Al(OH)_3$ and NaOH were supplied by Sigma Aldrich, H_2O_2 which acts as foaming agent was obtained by Merck, and hematite-derived from hematitation of mill scale industrial waste supplied by Research Center for Advanced Material – National Research and Innovation Agency (BRIN) Indonesia.

Fabrication of Hematite/Geopolymer Membrane via Direct Foaming Approach

Prior to utilizing fly ash as a raw material for membrane synthesis, it underwent characterization using XRF PanAlytical Minipal 4, XRD PanAlytical Expert Pro, and Particle Size Analyzer (PSA) Cilas 1090. The synthesis of a composite material consisting of hematite and geopolymer membrane was conducted using a method similar to the one described by Faradilla et al. [7]. Initially, a mixture was prepared by combining 65 g of fly ash from Pulang Pisau's Power Plant and 0.85 g of $Al(OH)_3$ with a base activator. The mixture was then stirred for 5 minutes until it was evenly mixed. Next, 3 g of hematite was added to the mixture, followed by the addition of H_2O_2 with different concentrations of 0, 2, 4, and 6 wt.% (referred to as FGH0, FGH2, FGH4, and FGH6). The mixture was stirred for an additional 2 minutes until it became homogeneous and formed a bubble. The paste mixture was put into the mould, which had a diameter of 4.5 cm and a height of 0.5 cm and left to cure for a duration of 7 days.

Membrane Characterization

The membrane conducted characterization using XRD PanAlytical Expert Pro for determining crystallography property, FTIR Shimadzu 8400S was carried out to observing functional group of membrane, SEM FEI Inspect S50 was used for capturing surface morphology of developed membrane, water contact angle test was conducted to determining surface wettability, and the Archimedes method to ascertain its porosity. The procedure for calculating Archimedes's technique is as Eq. 1 [7].

$$Porosity\ (\%) = \frac{W-D}{V} \times 100\%$$ (1)

Where, W is saturated cored weight, D is dry grain weight, and V is bulk volume.

Results and Discussion

This section was subdivided into multiple subsections. The first subsection discussed the properties of fly ash, while the second subsection focus on the structural analysis of the developed membrane. The final part of this section includes details regarding the membrane's porosity, morphology, and surface wettability

Composite Materials: SEAJCCM2024
Materials Research Proceedings 56 (2025) 1-10

Materials Research Forum LLC
https://doi.org/10.21741/9781644903636-1

Fly Ash Characteristics

Fly ash is mostly composed of silicon oxide (SiO_2), aluminum oxide (Al_2O_3), iron oxide (Fe_2O_3), and calcium oxide (CaO). The major compound of fly ash includes minor components such as magnesium, sodium, potassium, sulfur, and titanium oxides. The chemical composition of fly ash from Pulang Pisau's Power Plant is shown in Table 1. The analysis revealed that the fly ash consists of SiO_2, Al_2O_3, Fe_2O_3, and CaO in proportions of 29.00%, 9.98%, 13.75%, and 28.37%, respectively. According to Alterary et al. [20], fly ash can be categorized as C class if the amount of CaO exceeds 20% and the combined amount of SiO_2, Al_2O_3, and Fe_2O_3 exceeds 50%. Fly ash, due to its composition, can potentially be utilized as a raw material for creating geopolymer membrane materials.

Table 1. Major components of fly ash from Pulang Pisau's power plant.

Compound	Percentage (%)
SiO_2	29.00
Al_2O_3	9.98
Fe_2O_3	13.75
CaO	28.37

The size distribution of fly ash significantly influences the mechanical properties of the geopolymer membrane. Fly ash, characterized by its small particle size, can yield favorable mechanical qualities. Nath et al. [21] showed that an increase in particle size of fly ash results in decreased solubility in the base activator. Conversely, fly ash, characterized by its fine particle size, can easily dissolve and blend with the base activator. Therefore, it is a significant determinant that impacted the development of the geopolymer structure and its physical characteristics. The PSA measurement findings are presented in Table 2, while the particle size distribution is displayed in Fig. 1. The results indicate that the fly ash from Pulang Pisau's power plant has an average particle size of 30.34 µm. The cumulative diameter distribution shows that 10% of the particles have a diameter of 1.82 µm, 50% have a diameter of 24.06µm, and 90% have a diameter of 66.00 µm.

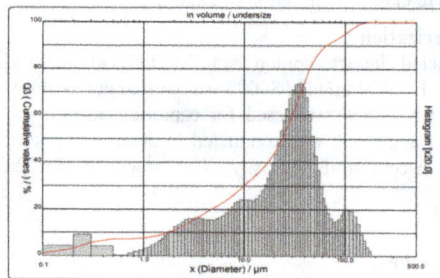

Fig. 1. Distribution of particle size of fly ash.

Table 2. Particle size analysis result.

Parameter	Particle Size (µm)
Diameter 10%	1.82
Diameter 50%	24.06
Diameter 90%	66.00

Composite Materials: SEAJCCM2024
Materials Research Proceedings 56 (2025) 1-10

Materials Research Forum LLC
https://doi.org/10.21741/9781644903636-1

Structural Characterization of Hematite/Geopolymer Membrane

Structural analysis is a crucial method for determining geopolymer formation. The diffractogram of fly ash from Pulang Pisau's power plant is shown in Fig. 2(a). The XRD patterns indicate that fly ash is primarily composed of intense quartz peaks at $2\theta=20.82°$ and $26.61°$, followed by mullite minerals diffraction at $2\theta = 31.2°$, $33.1°$, $35.4°$, $39.2°$, and $59.8°$. The characterization findings of geopolymer membranes and hematite/geopolymer membranes are presented in Fig. 2(b). The diffractogram indicates variations in both fly ash precursors and products (geopolymers). The diffractogram of fly ash shows the presence of quartz and mullite phases. The intensity of the peaks at $2\theta=20.82°$ and $26.61°$ decreases, indicating the transformation of the quartz phase into an aluminosilicate material, such as geopolymer. This transformation is further supported by the formation of an amorphous phase from mullite at 2θ around $40°$, which suggests the presence of a geopolymer network [21]. Furthermore, the diffraction pattern of the hematite/geopolymer membrane exhibits the presence of a hematite peak at 2θ angles of $33.4°$ and $64.7°$.

The FTIR characterization involved the use of a geopolymer membrane and a Hematite/Geopolymer membrane with 6 wt.% H_2O_2 as a pore-forming agent. Figure 3 displays the FTIR spectra, while Table 3 presents the representation of molecular vibration. The presence of T-O-T groups (where T represents Si or Al) in the aluminosilicate backbone of the geopolymer is indicated by the vibration at wavenumbers about 973 and 1345 cm^{-1}. The molecular vibration below 3500 cm^{-1} indicates the presence of the hydroxyl group (O-H) in the geopolymer network. The hydroxyl vibration of geopolymer membrane was broader than FGH6, it might be the presence of rich –OH from Si-OH or Al-OH from geopolymer backbone. On the other hand, the presence of hematite in geopolymer composite leads to reduce the vibration of hydroxyl group due to interaction of hematite with geopolymer backbone. The vibration at 1650 cm^{-1} is likely caused by H-O-H (H_2O) molecules that are physically trapped during geopolymerization. This finding is consistent with the previous research conducted by Yahya et al. [22].

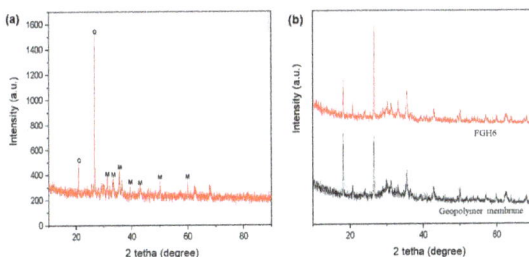

Fig. 2. Diffractogram of (a) fly ash, (b) geopolymer membrane and hematite/geopolymer membrane (FGH6).

Table 3. Molecular vibration of geopolymer and hematite/geopolymer membrane.

Wavenumber (cm^{-1})		Molecular Vibration	Ref
Geopolymer Membrane	Hematite/Geopolymer Membrane (FGH6)		
3459	3450	O-H	[22]
1651	1648	H-O-H	[22]
1345	1343	T-O-T	[21]
973	968	T-O-T	[21]

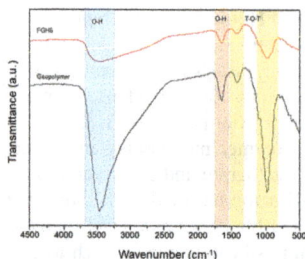

Fig. 3. FTIR Spectra of geopolymer and hematite/geopolymer membrane (FGH6).

Membrane Porosity, Morphology and Surface Wettability

Porosity is another crucial feature that significantly influences membrane separation. The elevation in H_2O_2 concentration during the production of the membrane resulted in an augmented membrane porosity, as depicted in Fig. 4. The inclusion of 6 wt.% H_2O_2 resulted in a significantly increased porosity of nearly 50% in the membrane, while the addition of 4 wt.% H_2O_2 led to a porosity of close to 40%. This outcome indicates that a greater quantity of H_2O_2 obtain in the generation of bubbles during geopolymerization, subsequently leading to the production of a larger number of membrane pores [23].

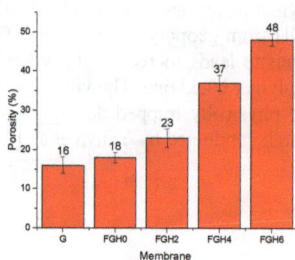

Fig. 4. Membrane porosity (n = 3).

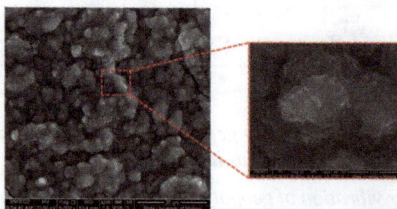

Fig. 5. SEM Micrograph of Geopolymer Membrane.

The porosity results are further corroborated by the morphology observation, which reveals the presence of microcracks on the top surface of the membrane containing 6 wt.% H_2O_2. Figure 5 depicts the structure of the geopolymer membrane, whereas Figure 6 illustrates the surface of the hematite/geopolymer membrane with the inclusion of 6 wt.% H_2O_2. According to the data from both Figures, the hematite/geopolymer membrane exhibited a higher number of microcracks as a result of the direct foaming process. On the other hand, the geopolymer membrane without the

Composite Materials: SEAJCCM2024   Materials Research Forum LLC
Materials Research Proceedings 56 (2025) 1-10   https://doi.org/10.21741/9781644903636-1

addition of H_2O_2 showed only a small number of microcracks. The presence of microcracks in membranes is not sufficient to rule them out for all applications; it depends on the operational requirements. Microcracks may not negatively impact performance in applications with low mechanical stress, such as passive filtration or low-pressure separation, as long as cracks do not create unfavourable paths for fluid or pollutants. In applications with non-critical selectivity, like coarse pre-filtration, the microcracks may not affect membrane performance. In contrast, the hematite/geopolymer membrane exhibited a rough surface with uneven forms in certain areas. The uneven shape of the particles on the membrane surface may be caused by the creation of pores during foaming. The particle on the surface of the geopolymer membrane exhibited a spherical shape, which differed from the original shape of a fly ash. However, when hematite was added as a composite, the surface structure changed significantly.

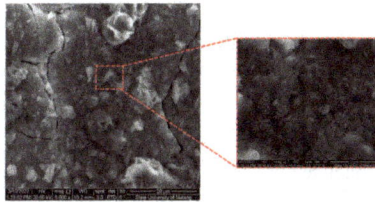

Fig. 6. Surface morphology of hematite/geopolymer membrane (FGH6).

Surface wettability is a significant characteristic in the field of membrane research. It may be assessed by measuring the water contact angle, which helps identify whether the surface is hydrophilic (water-attracting) or hydrophobic (water-repelling) [24]. This study investigated the water contact angle of a developed membrane to assess its wettability towards water. Figure 7(a) demonstrates that the geopolymer-based membrane had hydrophilic characteristics, as evidenced by water contact angle (WCA) data measuring below 90°. The unmodified geopolymer membrane displayed a WCA of 22.34°. However, the introduction of hematite effectively decreased the WCA value to 19.21°. The hydrophilicity of the hematite/geopolymer membrane was increased by including hydrophilic factors, indicating an improvement in the attraction of water molecules to the membrane surface. In addition, the use of a foaming agent during the manufacturing process results in a decrease in the WCA value. This discovery demonstrated a decline in trend as the concentration of H_2O_2 increased. As stated in the porosity results, an increase in porosity led to a greater occurrence of formation foaming during the geopolymerization of fly ash. This, in turn, contributed to an enhanced membrane porosity. As a result, a membrane with higher porosity causes a decrease in the WCA due to water being absorbed into the membrane.

Surface free energy (SFE) is a measure of the disruption of intermolecular interactions that occurs when a surface is formed. SFE can also be known to as interfacial free energy. The calculation of SFE can be carried out with the young equation [25]. The amount of SFE exhibited an inverse relationship with the WCA, whereby a lower WCA value corresponded to a higher SFE. The membrane surface's excellent wettability produced the highest SFE. FGH6 possesses a highest SFE value of 70.5 mJ/m², while the lower concentration of H_2O_2 resulted in a decrease in SFE. Subsequently, the unmodified membrane exhibits the lowest SFE value as a result of having the highest WCA value. Membrane porosity, WCA, and SFE are interrelated and affect many aspects of the membrane.

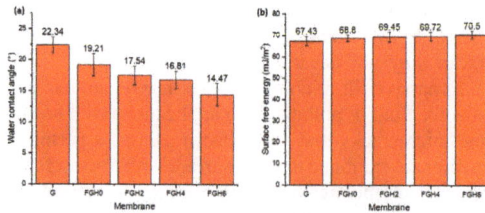

Fig. 7. Surface properties of membrane: (a) water contact angle/wettability and (b) surface free energy (n = 3).

Conclusions

The fly ash type C obtained from Pulang Pisau's Power Plant was effectively converted into a geopolymer and hematite/geopolymer membrane composite using the direct foaming process. The presence of T-O-T vibration at 1345 and 973 cm^{-1} confirms the existence of quartz phase, hematite, and amorphous phase in the diffraction patterns, indicating the creation of geopolymer. The direct foaming approach enhanced the porosity of the membrane by 50%. The SEM images reveal that the membrane surface has a rough texture and contains a limited number of microcracks as a result of the direct foaming process. The particle on the surface of the geopolymer membrane exhibited a spherical shape, which differed from the original shape of the fly ash. This indicates that the fly ash was effectively converted into a membrane composed of geopolymer and a composite of hematite/geopolymer.

Acknowledgements

The authors thank the Kurita Water and Environment Foundation (KWEF) Japan for providing Research Grant under Kurita Overseas Research Grant 2022 scheme with contract number 22pid023-K1. We also thank the Pulang Pisau's Power Plant of Central Kalimantan Indonesia for supplying the fly ash and the Research Center for Advanced Materials of National Research and Innovation Agency (BRIN) Indonesia for providing the hematite-derived from mill scale sample.

Credit authors contributions

Conceptualization, R.M.I., R.A, D.S.K.; data curation, E.R.S, R.M.I.; formal analysis, R.M.I., D.S.K., and D.S.K.; funding acquisition, R.M.I., R.A, D.S.K.; investigation, E.R.S.; methodology, R.M.I, H.F., and D.S.K.; validation, H.F, M.H.D.O, M.A.M.; visualization, R.A.; writing - original draft, R.M.I; writing – review and editing, R.M.I, D.S.K, M.E. All authors have read and agreed to publish a version of the manuscript.

Declaration of Competing Interests

The authors declare that they have no known competing financial interests or personal relationships that could have appeared to influence the work reported in this paper.

References

[1] A. Abdullah, K. Hussin, M.M.A. Abdullah, Z. Yahya, W. Sochacki, R.A. Razak, K. Bloch, H. Fansuri, The effects of various concentrations of NaOH on the inter-particle gelation of a fly ash geopolymer aggregate, Materials 14 (2021) 1111. https://doi.org/10.3390/ma14051111

[2] D.A.P. Wardani, L. Rosmainar, R.M. Iqbal, S.N. Simarmata, Synthesis and characterization of magnetic adsorbent based on Fe$_2$O$_3$-fly ash from Pulang Pisau's power plant of Central Kalimantan, IOP Conference Series: Materials Science and Engineering 980 (2020) 012014. https://doi.org/10.1088/1757-899X/980/1/012014

[3] I. Fatimah, P.W. Citradewi, R.M. Iqbal, S.A.I.S. Ghazali, A. Yahya, G. Purwiandono, Geopolymer from tin mining tailings waste using Salacca leaves ash as activator for dyes and peat water adsorption, S Afr J Chem Eng. 43 (2023) 257–65. https://doi.org/10.1016/j.sajce.2022.11.008

[4] M. Saukani, A.N. Lisdawati, H. Irawan, R.M. Iqbal, D.M. Nurjaya, S. Astutiningsih, Effect of nano-zirconia addition on mechanical properties of metakaolin-based geopolymer, J Compos Sci. 6 (2022) 293. https://doi.org/10.3390/jcs6100293

[5] S. Singh, S.K. Sharma, M.A. Akbar, Developing zero carbon emission pavements with geopolymer concrete: A comprehensive review, Transp Res D Transp Environ. 110 (2022) 103436. https://doi.org/10.1016/j.trd.2022.103436

[6] H. Fansuri, E. Erviana, M. Rosyidah, R.M. Iqbal, W.P. Utomo, Nurlina. Immobilization of chromium from liquid waste of electroplating home-industries by fly ash geopolymerization, IOP Conference Series: Materials Science and Engineering 367 (2018) 012049. https://doi.org/10.1088/1757-899X/367/1/012049

[7] F.S. Faradilla, Membran geopolimer berbahan dasar metakaolin yang dibuat dengan penambahan H_2O_2 secara direct foaming untuk pemisahan metilen biru, Doctoral dissertation, Institut Teknologi Sepuluh Nopember, Surabaya, 2021.

[8] H. Chen, Y.J. Zhang, P.Y. He, C.J. Li, H. Li, Coupling of self-supporting geopolymer membrane with intercepted Cr(III) for dye wastewater treatment by hybrid photocatalysis and membrane separation, Appl Surf Sci. 515 (2020) 146024. https://doi.org/10.1016/j.apsusc.2020.146024

[9] A. Naveed, Noor-Ul-Amin, F. Saeed, M. Khraisheh, M. al Bakri, S. Gul, Synthesis and characterization of fly ash based geopolymeric membrane for produced water treatment, Desalination Water Treat. 161 (2019) 126–31. https://doi.org/10.5004/dwt.2019.24283

[10] S. Subaer, H. Fansuri, A. Haris, M. Misdayanti, I. Ramadhan, T. Wibawa, Y. Putri, H. Ismayanti, A. Setiawan, Pervaporation membranes for seawater desalination based on geo–rGO–TiO₂ nanocomposites: Part 2—Membranes performances, Membranes (Basel) 12 (2022) 1046. https://doi.org/10.3390/membranes12111046

[11] K.K. Katibi, K.F. Md Yunos, H.C. Man, A.Z. Aris, M.Z.M. Nor, R.S. Azis, Influence of functionalized hematite nanoparticles as a reinforcer for composite PVDF-PEG membrane for BPF rejection: permeability and anti-fouling studies, Journal of Polymers and the Environment 31 (2023) 768-790. https://doi.org/10.1007/s10924-022-02605-z

[12] A. Salama, R. Abouzeid, W.S. Leong, J. Jeevanandam, P. Samyn, A. Dufresne, M. Bechelany, A. Barhoum, Nanocellulose-based materials for water treatment: Adsorption, photocatalytic degradation, disinfection, antifouling, and nanofiltration, Nanomaterials 11 (2021) 3008. https://doi.org/10.3390/nano11113008

[13] D.E. Fouad, C. Zhang, H. El-Didamony, L. Yingnan, T.D. Mekuria, A.H. Shah, Improved size, morphology and crystallinity of hematite (α-Fe₂O₃) nanoparticles synthesized via the precipitation route using ferric sulfate precursor, Results Phys. 12 (2019) 1253–1261. https://doi.org/10.1016/j.rinp.2019.01.005

[14] H. Wan, L. Hu, X. Liu, Y. Zhang, G. Chen, N. Zhang, R. Ma, Advanced hematite nanomaterials for newly emerging applications, Chemical Science 14.11 (2023) 2776-2798. https://doi.org/10.1039/d3sc00180f

Composite Materials: SEAJCCM2024
Materials Research Proceedings 56 (2025) 1-10

Materials Research Forum LLC
https://doi.org/10.21741/9781644903636-1

[15] D.S. Khaerudini, D.R. Insiyanda, Feasibility study of magnetite extracted from Indonesian mill scale through direct reduction promoted by graphite-based carbon, Journal of Chemical Technology & Metallurgy 55 (2020).

[16] D.S. Khaerudini, I. Chanif, D.R. Insiyanda, F. Destyorini, S. Alva, A. Pramono, Preparation and characterization of mill scale industrial waste reduced by biomass-based carbon, J Sustain Metall. 5 (2019) 510–518. https://doi.org/10.1007/s40831-019-00241-x

[17] N. Yong-Sing, L. Yun-Ming, H. Cheng-Yong, M.M.A.B. Abdullah, P. Pakawanit, P. Vizureanu, M.S. Khalid, N. Hui-Teng, H. Yong-Jie, M. Nabiałek, P. Pietrusiewicz, S. Garus, W. Sochacki, A. Śliwa, Improvements of flexural properties and thermal performance in thin geopolymer based on fly ash and ladle furnace slag using borax decahydrates, Materials 15 (2022) 4178. https://doi.org/10.3390/ma15124178 h

[18] N. Hui-Teng, H. Cheng-Yong, L. Yun-Ming, M.M.A.B. Abdullah, C. Rojviriya, H.M. Razi, S. Garus, M. Nabiałek, W. Sochacki, I.M.Z. Abidin, N. Yong-Sing, A. Śliwa, A.V. Sandu, Preparation of fly ash-ladle furnace slag blended geopolymer foam via pre-foaming method with polyoxyethylene alkyether sulphate incorporation, Materials 15 (2022) 4085. https://doi.org/10.3390/ma15124085

[19] F. Xia, S. Cui, X. Pu, Performance study of foam ceramics prepared by direct foaming method using red mud and K-feldspar washed waste, Ceram Int. 48 (2022) 5197–203. https://doi.org/10.1016/j.ceramint.2021.11.059

[20] S.S. Alterary, N.H. Marei, Fly ash properties, characterization, and applications: A review, J King Saud Univ. 33 (2021) 1–8. https://doi.org/10.1016/j.jksus.2021.101536

[21] S.K. Nath, S. Kumar, Role of particle fineness on engineering properties and microstructure of fly ash derived geopolymer, Constr Build Mater. 233 (2019) 1–9. https://doi.org/10.1016/j.conbuildmat.2019.117294

[22] Z. Yahya, M.M.A. Abdullah, K. Hussin, K.N. Ismail, R.A. Razak, A.V. Sandu, Effect of solids-to-liquids, Na_2SiO_3-to-NaOH and curing temperature on the palm oil boiler ash (Si+Ca) geopolymerisation system, Materials 8 (2015) 2227–42. https://doi.org/10.3390/ma8052227

[23] D. Yan, Y. Shi, Y. Zhang, W. Wang, H. Qian, S. Chen, Y. Liu, S. Ruan, A comparative study of porous geopolymers synthesized by pre-foaming and H_2O_2 foaming methods: Strength and pore structure characteristics, Ceram Int. 50 (2024) 17807–17. https://doi.org/10.1016/j.ceramint.2024.02.270

[24] M.F. Ismail, M.A. Islam, B. Khorshidi, A. Tehrani-Bagha, M. Sadrzadeh, Surface characterization of thin-film composite membranes using contact angle technique: Review of quantification strategies and applications, Adv Colloid Interface Sci. 299 (2022) 102524. https://doi.org/10.1016/j.cis.2021.102524

[25] X. Wang, Q. Zhang, Role of surface roughness in the wettability, surface energy and flotation kinetics of calcite, Powder Technol. 371 (2020) 55–63. https://doi.org/10.1016/j.powtec.2020.05.081

Composite Materials: SEAJCCM2024
Materials Research Proceedings 56 (2025) 11-20

Materials Research Forum LLC
https://doi.org/10.21741/9781644903636-2

Antioxidative Properties of Harumanis Mango Leaves using Microwave-Assisted Extraction

Nurfitrah Syahirah MOHD ASRI[1,a], Farizul Hafiz KASIM[1,2,b],
Noor-Soffalina SOFFIAN-SENG[3,c] and Khairul Farihan KASIM[1,2,d] *

[1]Faculty of Chemical Engineering & Technology, Universiti Malaysia Perlis, Arau 02600, Perlis, Malaysia

[2]Centre of Excellence for Biomass Utilization, Universiti Malaysia Perlis (UniMAP), Arau 02600, Perlis, Malaysia

[3]Department of Food Science, Faculty of Science and Technology, Universiti Kebangsaan Malaysia, 43600 UKM Bangi, Selangor, Malaysia

[a]syahirahfitrah@studentmail.unimap.edu.my, [b]farizul@unimap.edu.my, [c]soffalina@ukm.edu.my, [d]khairulfarihan@unimap.edu.my

Keywords: Antioxidant Activities, Mango Leaves, Microwave-Assisted Extraction, Harumanis Mango, Extraction

Abstract. In this study, a microwave-assisted extraction technique (MAE) was used to extract antioxidant compounds from the pruned leaves of Harumanis mango with water as the solvent. The one-factor-at-a-time (OFAT) approach was used to study the effect of several important parameters in the extraction process on antioxidant activity. These parameters include particle size (200, 100, 60, 30 and <30 mesh), sample size (0.5, 1.0, 1.5, 2.0, 2.5, and 3.0 g), and sample-to-solid ratio (2:50, 2:100, 2:150, 2:200, and 2:250 g/mL). Antioxidant activity was determined using the 1,1-diphenyl-2-picrylhydrazyl (DPPH) assay, the 2,2'-azino-bis(3-ethylbenzothiazoline-6-sulfonic acid) assay (ABTS), the reducing antioxidant power assay of iron (FRAP), and the reducing power (RP) assay. The Harumanis mango leaves were found to have high antioxidant activity in all the assays tested. The optimal parameters for extraction of antioxidant properties were a particle size of 200 mesh, a sample size of 2 g, and a sample-to-solvent ratio of 2:50 (g/mL). This set of parameters resulted in the most favourable results in the DPPH, ABTS, FRAP, and RP assays. The highest values observed in the DPPH, ABTS, FRAP and RP tests are 74.9%, 95.2%, 196.6 mg/mL and 797.2 mg/mL, respectively. Based on the R2 of the kinetics model tested, power law model was the most suitable to explain the variations of the extraction process of Harumanis mango leaves using MAE.

Introduction

Mangoes are scientifically known as *Mangifera indica* and belong to the genus Mangifera. This genus includes about 30 species of tropical fruit trees in the plant family Anacardiaceae [1]. In Malaysia, there are several varieties of mangoes that are cultivated. Well known varieties include Golek (MA 162), Masmuda (MA 204), Maha 65 (MA 165), Chok Anan (MA 224), Nam Dok Mai (MA 223), Alphonso (MA 173), and Harumanis (MA 128) [2]. According to the Department of Agriculture of Malaysia, the Harumanis is a popular variety grown in Malaysia and it is well sought after in the commercial market. The mango thrives particularly well in the state of Perlis, which is located in northern Malaysia.

Mango leaves are rich in phytochemical compounds such as phenolic acids, gallic acids, flavonoids, ascorbic acid, carotenoids, gallotannins, ellagic acid, mangiferin, quercetin, isoquercetin, β-glucogallinbenzophenones, and tocopherols among others [3]. The phytochemical compounds in mango leaves exhibit a wide range of bioactivity, including antioxidative,

Content from this work may be used under the terms of the Creative Commons Attribution 3.0 license. Any further distribution of this work must maintain attribution to the author(s) and the title of the work, journal citation and DOI. Published under license by Materials Research Forum LLC.

antidiabetic, antipyretic, antibacterial, anti-inflammatory, immunomodulatory, and analgesic activity [4-6]. Mango leaves have been found to have significant antioxidant activity due to the presence of phenols and flavonoids. Antioxidants, especially those that are extracted from natural sources have recently attracted much attention due to their exceptional ability to scavenge free radicals and reactive oxygen species [7]. Mango leaf extract is a promising antioxidant compound and has many potential applications in the food, packaging and many other industries [8].

There are several methods to extract antioxidant compounds from natural sources. Traditionally, these methods make use of organic solvents to dissolve the organic compounds of interest from the plant matrix, and this causes a lot of environmental and health problems. Therefore, the use of green extraction techniques such as microwave-assisted extraction (MAE) to extract antioxidants are becoming increasingly popular because they minimize the amount of organic solvent required and this in turn, reduces the amount of waste generated by the extraction process [9]. The main objectives of green extraction processes are to achieve fast extraction, efficient energy use, higher mass and heat transfer, while requiring fewer apparatus usage and processing steps. This study presents the optimal parameters to maximize the extraction of antioxidants from the leaves of Harumanis mango using MAE.

Kinetic modelling plays a vital role in understanding the extraction processes of bioactive compounds, such as antioxidants, from natural sources. In the context of Harumanis mango leaves, kinetic modelling can provide valuable insights into the rate of extraction and the efficiency of MAE. By applying mathematical models, it is possible to describe the dynamics of the extraction process, predict optimal conditions and understand the influence of key factors such as microwave power, extraction time and solvent concentration. Common kinetic models used in such studies include power law model, Peleg's model, first order model and Gaussian model. These models enable the estimation of thermodynamic properties and the activation energy, which are crucial for scaling up the extraction process in industrial applications.

Materials and Methods

Chemicals and materials

Ethanol and 2,2-diphenyl-1-1picrylhydrazy (DPPH) were obtained from Fisher Scientific Limited. 2,2'-azino-bis-ethylbenzothiazoline-6-sulfonic acid, potassium persulfate and methanol were purchased from HmbG Chemicals. Sodium acetate, 2,4,6-tri(2-pyridyl)-1,3,5-triazine (TPTZ) and ferric chloride were obtained from Sigma Aldrich Malaysia. All chemicals are analytical grade. The pruned young leaves of the Harumanis plant were collected from the Harumanis Plantation at Sungai Batu Pahat in Perlis, Malaysia.

Sample Preparation

The young leaves were carefully washed under running tap water to eliminate any dirt or contaminants. Afterward, the leaves were dried overnight in an oven (Thermo-Line SOV140B, China) at 55 °C. Once dried, the leaves were ground in a fine powder using an electric coffee grinder (Saachi Coffee Grinder, China). The powdered material was then sieved into five different particle sizes (200, 100, 60, 30 and <30 mesh) following the B.S.S. standard. The resulting samples were stirred in a freezer (Midea R600a, China) at -20 °C until needed for further analysis.

Microwave-Assisted Extraction

The antioxidant compounds from Harumanis leaves were extracted using a modified microwave oven system. This system equipped with a digital control unit, allows for precise regulation of irradiation time (ranging from 1 sec to 99 min) and microwave radiation power (from 10 to 900 W). The microwave oven (Sharp R357EK, China) was modified to include a condenser, ensuring solvent vapor recovery during heating and preventing the sample from drying out. Extraction was performed using water as the solvent under various conditions. In brief, the leaf samples were placed in a round-bottom flask with water and heated in the microwave while being stirred. After irradiation, the extracts were quickly cooled in an ice bath, then filtered to recover the antioxidant compounds, which were then stored in a freezer until analysis.

Screening of MAE Parameters using One-Factor-at-a-Time (OFAT) Approach

The extraction of antioxidant compounds from Harumanis mango leaves using MAE was optimized by screening various parameters through the One-Factor-at-a-Time (OFAT) method. The parameters evaluated included particle size (200, 100, 60, 30 and <30 mesh), sample size (0.5, 1.0, 1.5, 2.0, 2.5 and 3.0 g) and sample-to-solvent ratio (2:50, 2:100, 2:150, 2:200 and 2:250 w/v). Each parameter was adjusted individually, and the optimal condition determined for each was used in subsequent experiments.

Effect of Particle Size

The effect of particle size on the antioxidant properties was determined using 2 g of sample and 50 mL of water at a microwave radiation power of 800 W and constant heating duration of 3 min. The liquid extract was cooled to room temperature and then filtered to recover the antioxidant compounds. The antioxidant compounds were stored in the freezer at -20°C until further analysis.

Effect of Sample Size

After analysis of the results obtained from effect of particle size, a particle size of 200 mesh was chosen. The amount of solvent is fixed at 50 mL, the time duration is fixed at 3 min and microwave radiation is fixed at 800 W. The effect of sample size was determined by varying the sample size from 0.5, 1.0, 1.5, 2.0, 2.5 to 3.0 g, then measuring the antioxidative property of the recovered compounds after the extraction process.

Effect of Sample-to-Solvent Ratio

After analysis of the results obtained from effect of sample size, a particle size of 200 mesh and sample size of 2 g was chosen. The effect of sample-to-solvent ratio was investigated by varying the sample-to-solvent ratio from 2:50, 2:100, 2:150, 2:200 to 2:250 (w/v), then measuring the antioxidative property of the recovered compounds after the extraction process. Other parameters remained the same as described previously.

Effect of Microwave Radiation Power and Time on the Antioxidant Activities of Harumanis Leaves

This experiment was conducted using microwave radiation power of 80, 160, 240, 320, 400, 480, 560, 640, 720, and 800 W, and extraction duration of up to 10 minutes. The other parameters are a particle size of 200 mesh, a sample size of 2 g and solvent-to-sample ratio of 2:50 (w/v).

Antioxidant Assay

DPPH Assay

The antioxidant potential of the extract was assessed using the 1,1-diphenyl-2-picrylhydrazyl (DPPH) assay [10]. For the analysis, 200 µL of the extract was combined with 2.5 mL of 60 µM DPPH solution in ethanol and the mixture was stirred under dark conditions for 30 min. Distilled water was used as negative control. The absorbance of the sample was then recorded at 517 nm using a UV-VIS spectrophotometer (Thermo Spectronic Genesys 20, USA). DPPH radical scavenging activity was calculated using Eq. 1.

$$\text{DPPH scavenging activity (\%)} = (1 - \frac{\text{(Absorbance of sample at 517nm)}}{\text{(Absorbance of control at 517nm)}}) \times 100 \qquad (1)$$

ABTS Assay

The ABTS assay is considered a sensitive approach for measuring antioxidant activity due to the fast reaction kinetics of the antioxidants. The ABTS radical scavenging activity was

Composite Materials: SEAJCCM2024
Materials Research Proceedings 56 (2025) 11-20

Materials Research Forum LLC
https://doi.org/10.21741/9781644903636-2

determined according to the procedure described in Zhu et al. [11] with some modifications. In brief, a solution of 7 mM 2,2'-azino-bis(3-ethylbenzothiazoline-6-sulphonic acid) (ABTS) in distilled water is first prepared. The initial solution is then added with potassium persulfate, bringing the final concentration of potassium persulfate to 2.45 mM. 2.45 Mm potassium persulfate was used to transform the ABTS into its radical cationic form (ABTS·+), and the mixture was lest to remain at room temperature in the dark for 12 to 16 hours before to use. Methanol was used to dilute the ABTS·+ solution until the absorbance at 734 nm was 0.70 (±0.02). The reaction mixture consists of 0.07 mL sample extract and 3 mL ABTS radical. After incubation for 30 min, the absorbance was determined at 734 nm. The antioxidant activity was calculated using Eq. 2.

$$\text{Inhibition (\%)} = \left(\frac{(\text{Absorbance of control} - \text{Absorbance of sample})}{(\text{Absorbance of control})} \right) \times 100 \qquad (2)$$

Ferric Reducing Antioxidant Power (FRAP)
The ferric reducing antioxidant power (FRAP) assay was prepared using a method previously described in Shang et al. [12]. In brief, 0.3 M pH 3.6 sodium acetate buffer, 0.01 M of TPTZ solution and 0.02 M of ferric chloride solution were mixed at a volume ratio of 10:1:1 to obtain the FRAP working solution. Then, 200 µL diluted supernatant was added to 3.8 mL FRAP reaction solution for 4 min and its absorbance at 593 nm was measured. Known quantities of ferrous sulphate were added and the absorbance was measured to construct the standard curve. The results were expressed in milligram equivalents of $FeSO_4$ per milligram of dry weight.

Reducing Power (RP)
Aliquots of 1 mL of the extract was mixed with 2.5 mL of 1% potassium ferrocyanide and 2.5 mL of 0.2 mM pH 6.6 phosphate buffer. The mixture was heated to 50 °C in a water bath and allowed to react for 20 min. Then, 2.5 mL of 10% trichloroacetic acid was added followed by centrifugation at 3000 rpm for 10 min. Then, 2.5 mL of the supernatant was recovered and it was mixed with 2.5 mL of distilled water and 0.5 mL of 0.1% ferric chloride. Absorbance of the sample was then measured at a wavelength of 700 nm. The negative blank was prepared by substituting the diluted extract with the same amount of methanol.

Kinetic Modelling of the Harumanis Leaves Extraction
The data from the extraction condition were analysed by using various kinetic models including the first-order model, power law model, Gaussian model and Peleg's model. Kinetic extraction analysis of antioxidant activities was conducted using graphical methods. The graphs were fitted to the representative models using a Sigmaplot (Version 12), Systat (2013) software.

Statistical Analysis
The mean and standard deviation of all measurements were determined using Minitab® (version 19; StataCorp LLC, TX, USA).

Results and Discussion
Screening of the extraction parameters

Effect of Particle Size
The effect of particle size on the antioxidant activity of Harumanis mango leaf extract is shown in Fig. 1. The results show a consistent trend across all measured antioxidant activities, demonstrating that antioxidant efficacy is inversely proportional to particle size. Specifically, the DPPH and ABTS scavenging activities were high at 73.6% and 93.3%, respectively, and the FRAP and RP values reach their highest at 193.6 and 760.1 mg/mL, respectively when using the smallest particle size (200 mesh). The observed increased antioxidant yield when using a smaller particle size can be attributed to the higher surface-to-volume ratio which enhances the interaction between

Composite Materials: SEAJCCM2024
Materials Research Proceedings 56 (2025) 11-20

Materials Research Forum LLC
https://doi.org/10.21741/9781644903636-2

the solvent and the plant material during extraction. Particle size significantly impacts extraction efficiency by influencing mass transfer kinetics and solvent access to soluble compounds [13]. Notably, ABTS scavenging activity is always higher than 80% regardless of particle sizes, indicating that Harumanis mango leaves contain a substantial amount of antioxidants and have considerable antioxidant potential. The ABTS assay is recognized for its sensitivity and reliability in assessing antioxidant activity due to its rapid reaction kinetics and higher responsiveness to antioxidants compared to other radicals [14].

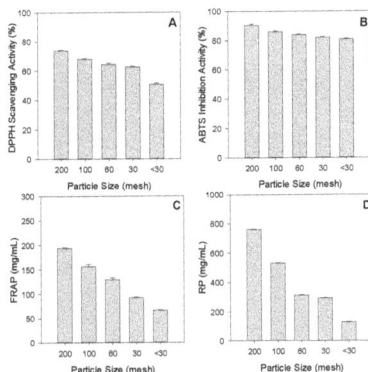

Fig. 1. Effects of particle size on the antioxidant activities; (A) DPPH scavenging activity, (B) ABTS inhibition activity, (C) FRAP content and (D) RP content.

Effect of Sample Size

Fig. 2 illustrates the effects of sample size on the antioxidant activities (DPPH, ABTS, FRAP and RP) of Harumanis mango leaves. The optimal sample size for extraction using MAE was found to be 2 g, with a particle size of 200 mesh and a solvent volume of 50 mL. Antioxidant activity increased with sample size, peaking at 2 g, after which it began to decrease. A larger sample size provides a greater surface area for interaction with the solvent, thus improving the extraction of antioxidant compounds from the leaves. However, beyond 2 g, an increase in sample size leads to insufficient solvent volume, which reduces the efficiency of the extraction process [15]. It is also important to note that finely ground samples offer a more uniform texture, enabling more consistent and effective extraction, whereas coarser samples tent to yield uneven extraction due to varying surface areas [16].

Effect of Sample-to-Solvent Ratio

The antioxidant activity was found to be inversely proportional to the sample-to-solvent ratio (Fig. 3). A ratio of 2:50 (w/v) was determined to be optimal for extracting antioxidants from Harumanis mango leaves, yielding DPPH and ABTS scavenging activities of 74.9% and 91.4%, respectively, and FRAP and RP values of 196.6 and 797.2 mg/mL, respectively. This result aligns with the mass transfer principle, which explains that mass transfer is driven by the concentration gradient between the sample and the solvent [17]. At higher sample-to-solvent ratios, the solvent becomes rapidly saturated, resulting in poorer extraction from the mango leaves [15].

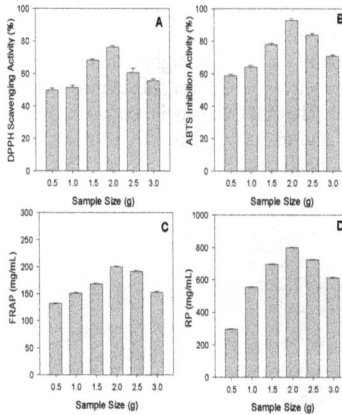

Fig. 2. Effects of sample size on the antioxidant activities; (A) DPPH scavenging activity, (B) ABTS inhibition activity, (C) FRAP content and (D) RP content.

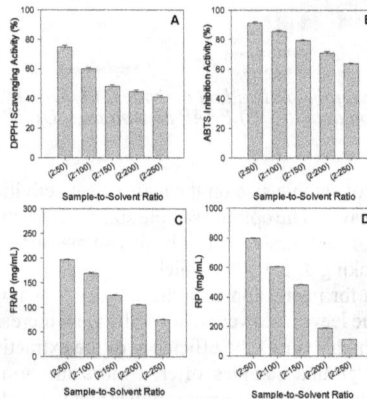

Fig. 3. Effects of sample-to-solvent ratio on the antioxidant activities; (A) DPPH scavenging activity, (B) ABTS inhibition activity, (C) FRAP content and (D) RP content.

Effect of Microwave Radiation Power and Time on the Antioxidant Activities of Harumanis Leaves

From the screening experiments, it can be concluded that the optimal parameters for extracting Harumanis mango leaves using MAE are a particle size of 200 mesh, a sample size of 2 g, and a solvent-to-sample ratio of 2:50 w/v. In MAE, microwave radiation power and extraction time are the primary factors that influence the effectiveness of antioxidant extraction. Therefore, this study further investigates the influence of microwave radiation power and time on the antioxidant activities of Harumanis mango leaf extract. The results showed that antioxidant activity changes depending on microwave power and heating duration (Fig. 4). DPPH scavenging activity, FRAP, and RP values generally increased with heating duration, whereas ABTS values appeared to decrease with heating time. For DPPH, FRAP, and RP, three distinct regions were observed based on the microwave radiation power level. In the first region (obtained when microwave power level ranged from 80 to 400 W), the antioxidant activity was found to increase consistently with heating duration. In the second region (obtained when microwave power level ranged from 480 to 560 W), there was a rapid initial increase in DPPH, FRAP and RP antioxidant activity in the first 5 to 6 min of extraction. Then, these antioxidant activities enter a degradation phase. This degradation suggests suboptimal extraction conditions, likely due to factors such as excessive extraction time or microwave power, which may adversely affect the stability of the antioxidants.

In the third region, (obtained when microwave power level ranged from 640 to 800 W was used), the DPPH, FRAP and RP antioxidant activity decreased with heating duration. DPPH antioxidant activity showed a steady decline with heating duration. FRAP and RP antioxidant activity initially increased after the first few minutes of MAE, then started to decrease after 2 min. This is because the high microwave power also damages the antioxidants in the extract. The finding reveals that the optimum extraction conditions to yield the optimum DPPH, FRAP, and RP activities were achieved at a microwave power of 480 W. For DPPH activity, the optimal extraction time was 5 min, yielding 85.8%. For FRAP activity, the optimal extraction time was 6 min, yielding a value of 198.17 mg/mL. Similarly, for RP activity, the optimal extraction time was also 6 min, yielding a value of 792.78 mg/mL. Interestingly, a previous study [18] also reported that microwave power has significant impact on antioxidant capacity. It was reported that the highest antioxidant activity of 15.82 mg TE/g (DPPH) from *Lentinula edodes* (shiitake) mushrooms extract was achieved at a microwave radiation power of 600 W and an extraction time of 15 min.

Kinetic Modelling of the Harumanis Leaves Extraction

A mathematical description of the process of extracting antioxidant activities from Harumanis mango leaves was carried out with the assistance of four extraction kinetic models that are extensively used which are power law model, Peleg's model, first order model and Gaussian model. The coefficient of determination (R^2) was employed as a creation for the selection of the most optimal mathematical model applicable to the extraction of Harumanis mango leaves. Table 1 provides a detailed presentation of the R^2 values for the analyses of antioxidant activities for mango leaves across all kinetic models. A parallel trend was observed where the power law model outperformed other models. The dominance of the power law model can be attributed to the fact that in the majority of analyses, it yielded the highest R^2 values. From the observation, the power law model emerges as the most effective for describing the relationship between ABTS, FRAP and RP. The power law model shows the highest R^2 value of 0.9917, indicating a very strong fit for RP. For DPPH, the best model was Gaussian model which provides the highest R^2 value of 0.9596. The FRAP activity demonstrates the highest R^2 value with the power law model, indicating a strong fit. Additionally, the ABTS activity also illustrates that the power law model shows the highest value R^2 of 0.9813, suggesting a very strong fit.

As the power law model emerged as the most suitable for accurately capturing the dynamics of antioxidant activities in this study, the model was applied to fit the curves generated at various levels of microwave power. In the power law model equation (Eq. 3), both coefficient, B and n served as parameters that could be adjusted. These parameters played a crucial role in refining the model. The flexibility of adjusting B and n allowed for a customized adaptation of the model, ensuring its ability to accurately represent the variations in the extraction process under different experimental conditions. It was noted that the microwave power exerted an influence on the parameter B. Specifically, B demonstrated an increment with the rise in microwave power up to 480 W. However, beyond this point, B exhibited a decline with further increases in microwave power up to 800 W. Enhanced microwave power was observed to accelerate molecular motion and internal diffusion, resulting in the breakdown of plant material. This, in turn, improved the solvents' ability to penetrate the plant matrix and minimized mass transfer barriers for internal diffusion from the solid phase to the liquid phase. Nevertheless, it is crucial to note that excessive microwave power might result in the absorption of too much energy by the samples, potentially causing degradation of the target compounds. The current study findings align with this observation, demonstrating that an increase in microwave power initially boosted the extraction yield up to a certain threshold, beyond which further increments yielded insignificant improvements.

$$C_t = B_t{}^n \qquad (3)$$

Where C_t is the extractable substance content at time t, n is the power law exponent (<1).

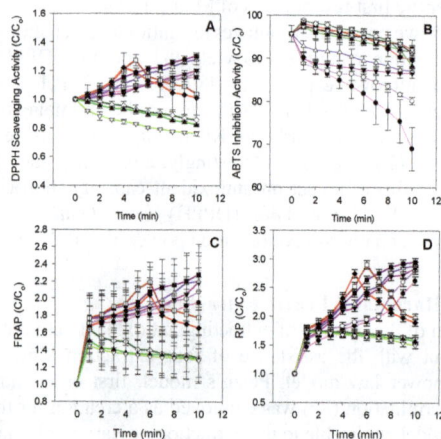

Fig. 4. Effect of microwave radiation power at 80 (—), 160 (—), 240 (—), 320 (—), 400 (—), 480 (—), 560 (—), 640 (—), 720 (—) and 800 W (—) over time on the antioxidant activities of Harumanis leaves extract; (A) DPPH radical scavenging activity, (B) ABTS inhibition activity, (C) FRAP content and (D) RP content.

Table 1. The coefficient of determination (R^2) for each model of antioxidant activities analyses.

Kinetics Model	Power Law	Peleg's	First Order	Gaussian
	R^2			
DPPH	0.9252 ± 2.1624^a	0.8247 ± 4.3409^a	0.9324 ± 1.4237	0.9596 ± 0.1119^b
ABTS	0.9813 ± 0.0897^a	0.6256 ± 4.1490^b	0.9658 ± 1.4869^b	0.9586 ± 1.0641^b
FRAP	0.9877 ± 0.0775^a	0.9849 ± 0.0654^a	0.9534 ± 0.0985^a	0.8102 ± 0.1452^b
RP	0.9917 ± 0.2110^a	0.8679 ± 0.3929^b	0.8671 ± 0.4108^b	0.8791 ± 0.1612^b

Conclusion

This study demonstrated the extraction of antioxidants from Harumanis mango leaves extracted using microwave-assisted extraction (MAE). Three parameters -particle size, sample size, and sample-to-solvent ratio- were evaluated using the OFAT method. It was found that the optimal extraction parameters are a particle size of 200 mesh, a sample size of 2 g, and a sample-to-solvent ratio of 2:50 (g/mL). The effect of microwave power level and heating duration were also investigated. The optimal microwave power level to use is 480 W. The required extraction time is 5 min for maximum DPPH and 6 min for maximum FRAP and RP. Future studies should explore the extraction mechanisms of the antioxidants and their interaction with other phytochemicals in Harumanis mango leaves.

Acknowledgement
The author would like to acknowledge the support provided by the Fundamental Research Grant Scheme (FRGS) under grant number FRGS/1/2020/TK0/UNIMAP/03/25 from the Ministry of Higher Education Malaysia and Universiti Malaysia Perlis for providing the equipment and facilities.

References

[1] C.Y. Huang, C.H. Kuo, C.H. Wu, A.W. Kuan, H.R. Guo, Y.H. Lin, P.K. Wang, Free Radical-Scavenging, Anti-Inflammatory, and Antibacterial Activities of Water and Ethanol Extracts Prepared from Compressional-Puffing Pretreated Mango (Mangifera indica L.) Peels, J. Food Qual. 1 (2018) 1025387. https://doi.org/10.1155/2018/1025387

[2] M.A. Sani, H. Abbas, M.N. Jaafar, M.B. Abd Ghaffar, Morphological characterisation of Harumanis mango (Mangifera indica Linn.) in Malaysia, Int. J. Environ. Agric. Res. 4 (2018) 45-51.

[3] M. Kumar, V. Saurabh, M. Tomar, M. Hasan, S. Changan, M. Sasi, C. Maheshwari, U. Prajapati, S. Singh, R.K. Prajapat, S. Dhumal, S. Punia, R. Amarowicz, M. Mekhemar, Mango (Mangifera indica L.) leaves: Nutritional composition, phytochemical profile, and health-promoting bioactivities, Antioxidants 10 (2021) 299. https://doi.org/10.3390/antiox10020299

[4] D. Alshammaa, Preliminary screening and phytochemical profile of Mangifera indica leave's extracts, cultivated in Iraq, Int. J. Curr. Microbiol. Appl. Sci. 5 (2016) 163-173. https://doi.org/10.20546/ijcmas.2016.509.018

[5] Y. Kabir, H.U. Shekhar, J.S. Sidhu, Phytochemical compounds in functional properties of mangoes, Handbook of mango fruit: Production, postharvest science, processing technology and nutrition, Wiley-Blackwell, United States, 2017, pp 237-254. https://doi.org/10.1002/9781119014362.ch12

[6] J. Pan, X. Yi, S. Zhang, J. Cheng, Y. Wang, C. Liu, X. He, Bioactive phenolics from mango leaves (Mangifera indica L.), Ind. Crop. Prod. 111 (2018) 400-406. https://doi.org/10.1016/j.indcrop.2017.10.057

[7] F. Giampieri, J.M. Alvarez-Suarez, S. Tulipani, A.M. Gonzàles-Paramàs, C. Santos-Buelga, S. Bompadre, J.L. Quiles, B. Mezzetti, M. Battino, Photoprotective potential of strawberry (Fragaria× ananassa) extract against UV-A irradiation damage on human fibroblasts, J. Agric. Food Chem. 60 (2012) 2322-2327. https://doi.org/10.1021/jf205065x

[8] Y. Kumar, V. Kumar, Sangeeta, Comparative antioxidant capacity of plant leaves and herbs with their antioxidative potential in meat system under accelerated oxidation conditions, J. Food Meas. Charact. 14 (2020) 3250-3262. https://doi.org/10.1007/s11694-020-00571-5

[9] A.H. Nour, A.R. Oluwaseun, A.H. Nour, M.S. Omer, N. Ahmed, Microwave-assisted extraction of bioactive compounds, In Microwave Heating-Electromagnetic Fields Causing Thermal and Non-Thermal Effects, IntechOpen, London, UK, 2021, pp 235-246. https://doi.org/10.5772/intechopen.96092

[10] S. Ngamsuk, T.C. Huang, J.L. Hsu, Determination of phenolic compounds, procyanidins, and antioxidant activity in processed Coffea arabica L. leaves, Foods 8 (2019) 389. https://doi.org/10.3390/foods8090389

Materials Research Forum LLC
https://doi.org/10.21741/9781644903636-2

[11] Z. Zhu, J. Chen, Y. Chen, Y. Ma, Q. Yang, Y. Fan, C. Fu, B. Limsila, R. Li, W. Liao, Extraction, structural characterization and antioxidant activity of turmeric polysaccharides, LWT. 154 (2022) 112805. https://doi.org/10.1016/j.lwt.2021.112805

[12] A. Shang, M. Luo, R.Y. Gan, X.Y. Xu, Y. Xia, H. Guo, Y. Liu, H.B. Li, Effects of microwave-assisted extraction conditions on antioxidant capacity of sweet tea (Lithocarpus polystachyus Rehd.), Antioxidants 9 (2020) 678. https://doi.org/10.3390/antiox9080678

[13] O.A. Pătrăuanu, L. Lazăr, V.I. Popa, I. Volf, Influence of particle size and size distribution on kinetic mechanism of spruce bark polyphenols extraction, Cellul. Chem. Technol. 53 (2019) 71–78. https://doi.org/10.35812/CelluloseChemTechnol.2019.53.08

[14] E.S. Prasedya, A. Frediansyah, N.W.R. Martyasari, B.K. Ilhami, A.S. Abidin, H. Padmi, Fahrurrozi, A.B. Juanssilfero, S. Widyastuti, A.L. Sunarwidhi, Effect of particle size on phytochemical composition and antioxidant properties of Sargassum cristaefolium ethanol extract, Sci. Rep. 11 (2021) 17876. https://doi.org/10.1038/s41598-021-95769-y

[15] O.R. Alara, N.H. Abdurahman, S.K.A. Mudalip, O.A. Olalere, Effects of microwave-assisted extraction parameters on the recovery yield and total phenolic content of Vernonia amygdalina leaf extracts with different methods of drying, Jundishapur J. Nat. Pharm. Prod. 14 (2019) 57620. https://doi.org/10.5812/jjnpp.57620

[16] Q.W. Zhang, L.G. Lin, W.C. Ye, Techniques for extraction and isolation of natural products: A comprehensive review, Chin. Med. 13 (2018) 1-26. https://doi.org/10.1186/S13020-018-0177-X

[17] I.S. Che Sulaiman, M. Basri, H.R. Fard-Masoumi, W.J. Chee, S.E. Ashari, M. Ismail, Effects of temperature, time, and solvent ratio on the extraction of phenolic compounds and the anti-radical activity of Clinacanthus nutans Lindau leaves by response surface methodology, Chem. Cent. J. 11 (2017) 1-11. https://doi.org/10.1186/s13065-017-0285-1

[18] W. Xiaokang, J.G. Lyng, N.P. Brunton, L. Cody, J.C. Jacquier, S.M. Harrison, K. Papoutsis, Monitoring the effect of different microwave extraction parameters on the recovery of polyphenols from shiitake mushrooms: Comparison with hot-water and organic-solvent extractions, Biotechnol. Rep. 27 (2020) e00504. https://doi.org/10.1016%2Fj.btre.2020.e00504

Composite Materials: SEAJCCM2024
Materials Research Proceedings 56 (2025) 21-29

Materials Research Forum LLC
https://doi.org/10.21741/9781644903636-3

A Review of Reactivity of Precursor in Manufacturing Non-Fired Clay Brick

Jun-Jian KOO[1,a], Mohammad Zawawi ROSMAN[2,b], Chee-Ming CHAN[2,c*],
Noor Khazanah A RAHMAN[2,d], Salina SANI[2,e] and Nur Faezah YAHYA[2,f]

[1]Soil Instruments (M) Sdn. Bhd., 12, Jalan Utarid U5/14, Seksyen U5, 40150 Shah Alam, Selangor, Malaysia

[2]Faculty of Engineering Technology, Universiti Tun Hussein Onn Malaysia, 84600 Pagoh, Johor, Malaysia

[a]gn230024@student.uthm.edu.my, [b]mzawawirosman@gmail.com, [c]chan@uthm.edu.my, [d]khazanah@uthm.edu.my, [e]salinas@uthm.edu.my, [f]nurfaezah@uthm.edu.my

Keywords: Geopolymerisation, Non-Fired Brick, Precursor, Strength, Reactivity

Abstract. Geopolymer is a product of aluminosilicate materials with a strongly alkaline solution mixture, curing it at a low or slightly high temperature. Because of its low energy consumption and low carbon emissions, clay brick is recognised as a green, sustainable construction material. It is formed by three main stages of the geopolymerisation process: the dissolution of aluminosilicate materials, the nucleation growth and polymerization of monomers, and the reorganisation and polycondensation of those monomers. However, there is one critical element that should be noted, which is the reactivity of the precursor, which will affect the overall geopolymerisation process and also the final product, geopolymer. There are four factors that will affect the reactivity of a precursor: chemical composition, glass phase content, morphology, and mineralogy. If there is an issue with the low reactivity of a precursor, there are three techniques, which are thermal, chemical, and mechanical treatment, to transform it into high reactivity properties. In conclusion, with respect to the manufacturing of geopolymer bricks, also called non-fired clay bricks, the properties of the precursor in terms of reactivity should be considered in order to achieve the desired properties in the construction field.

Introduction

Geopolymer is the third generation of cement instead of lime and ordinary Portland cement [1]. The 3D transition from amorphous to semi-crystalline geopolymer structure is formed by a chemical reaction between a silica- and alumina-rich precursor and a strongly alkaline medium at room temperature or elevated temperatures [2]. It is eco-friendly and able to be a solution to the issue of global warming due to its low carbon footprint. The manufacturing process is low in energy consumption, carbon emissions, and cost. [3,4]. Hence, it could be known as the "green materials" that are able to be used in the construction field, such as the production of brick. The geopolymer possesses outstanding mechanical and durability properties, such as high compressive strength, low water absorption, high resistance to acid and chemical solutions, and stability at high temperatures [5,6].

The final properties of geopolymer in terms of compressive strength are determined by the nature of the precursor, which can be called raw materials, especially the reactivity of the raw materials. The higher the reactivity, the higher the quality of the geopolymer formation. In addition, there are available techniques that can be applied to improve the reactivity of the raw materials, such as thermal, mechanical, and alkali treatment [7]. In this paper, the reactivity of the precursor is reviewed for the manufacturing of non-fired clay bricks with desired properties according to the standard of the construction application.

Content from this work may be used under the terms of the Creative Commons Attribution 3.0 license. Any further distribution of this work must maintain attribution to the author(s) and the title of the work, journal citation and DOI. Published under license by Materials Research Forum LLC.

Composite Materials: SEAJCCM2024   Materials Research Forum LLC
Materials Research Proceedings 56 (2025) 21-29   https://doi.org/10.21741/9781644903636-3

Geopolymerisation

Geopolymerisation is an exothermic process [8]. It is divided into two phases: reacting and curing as shown in Fig. 1 [9,10]. All reactions occur simultaneously while in the curing phase, with the settling and hardening reactions occurring at room temperature or higher to aid in the chemical reaction to improve the compressive strength of the geopolymer. Note that clay bricks are generally baked at elevated temperature of about 90°c to hasten the strength gain process.

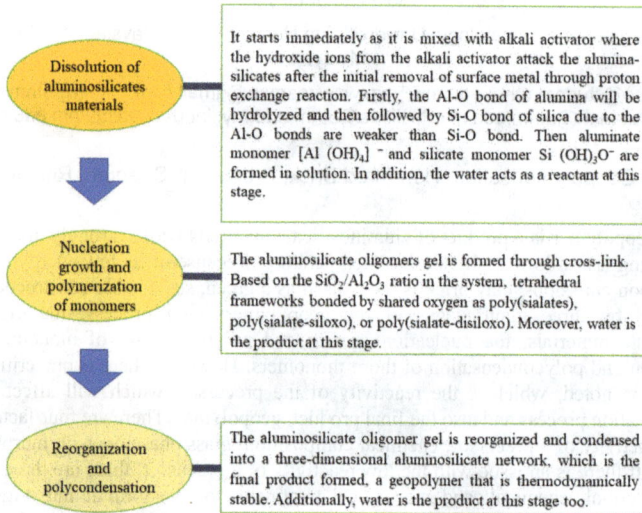

Dissolution of aluminosilicates materials	It starts immediately as it is mixed with alkali activator where the hydroxide ions from the alkali activator attack the alumina-silicates after the initial removal of surface metal through proton exchange reaction. Firstly, the Al-O bond of alumina will be hydrolyzed and then followed by Si-O bond of silica due to the Al-O bonds are weaker than Si-O bond. Then aluminate monomer $[Al (OH)_4]^-$ and silicate monomer $Si (OH)_3O^-$ are formed in solution. In addition, the water acts as a reactant at this stage.
Nucleation growth and polymerization of monomers	The aluminosilicate oligomers gel is formed through cross-link. Based on the SiO_2/Al_2O_3 ratio in the system, tetrahedral frameworks bonded by shared oxygen as poly(sialates), poly(sialate-siloxo), or poly(sialate-disiloxo). Moreover, water is the product at this stage.
Reorganization and polycondensation	The aluminosilicate oligomer gel is reorganized and condensed into a three-dimensional aluminosilicate network, which is the final product formed, a geopolymer that is thermodynamically stable. Additionally, water is the product at this stage too.

Fig. 1. Geopolymerisation process at reacting phase.

Monitoring of each geopolymerisation stage is crucial to ensure the production of high-quality and durable geopolymer-based bricks. Each stage has a distinctive role in determining the final properties of non-fired clay bricks, especially in terms of performance in the areas of strength, durability and environmental stability. In brief, the initial alkaline activation stage would determine the brick's viscosity, workability and early strength development. In the ensuing gel formation stage, the materials' early setting time and strength gain are influenced by polymerisation of the dissolved silica and alumina into a gel-like structure. The brick's dimensional stability, mechanical strength development as well as durability are primarily determined in the intermediate stage of polymerisation and hardening stage due to densification of the aluminosilicate network. The brick then undergoes curing for its final hardness and strength, which invariably affects its performance in terms of water absorption, thermal and fire resistance.

Factors Affecting Reactivity of the Precursor

A precursor's reactivity is affected by four factors: chemical composition, glass phase content, morphology, and mineralogy. These factors must be carefully examined to determine whether the precursors are appropriate for producing the product based on the applications.

Chemical Composition

The precursor for synthesising the geopolymer should be high in alumina and silica. Conventionally a high silica content of >40% is required, in combination with 10-20% of alumina. It can be supported by the empirical formula for poly(sialates) of geopolymer, which is Mn {-(SiO_2) z-AlO_2} n, wH_2O. The values represented by the letter "z" describe the Si/Al ratio, while

Composite Materials: SEAJCCM2024   Materials Research Forum LLC
Materials Research Proceedings 56 (2025) 21-29   https://doi.org/10.21741/9781644903636-3

the values represented by the letter "n" are the polycondensation degree, and the values represented by the letter "w" are the amounts of water. The chemical composition can be identified by the technique of X-ray fluorescence which has been applied to the metakaolin [11].

Generally, the precursor can be obtained from natural sources, which is the primary raw material, and from waste and by-products, which is the second raw material [2]. The raw material to be reviewed in this paper is clays, which can form a geopolymer that meets the above-stated requirement. There is some research on the manufacturing of geopolymer bricks according to the types of clays and also other materials that are rich in silica and alumina as the major content, as shown in Table 1, in terms of compressive strength and water absorption In short, low-quality bricks are produced by clay without calcination due to the low reactivity of clay with an alkali medium [12]. However, the others with thermal treatment, high compacting pressure, and the addition of highly reactive materials will result in high compressive strength and low water absorption.

Table 1. Characteristic of non-fired bricks with sodium-based stabilisers.

Type of raw materials	Compressive strength [MPa]	Water absorption [%]	Ref.
Clay	2.91	13.51	[3]
Sandy clay loam	1.46 at 28 days	20.50	[13]
Low iron lateritic clay (LC) and aluminum hydroxide fine powder	0.01	-	[14]
Calcinated LC and aluminum hydroxide fine powder (Al (OH)$_3$)	24.8	7.07	[14]
Clay (C) and Class-F Flay Ash (FA)	23	8	[15]

Glass Phase Content

The degree of structural order of the raw materials is inversely proportional to their reactivity. The geopolymerisation process is favoured in amorphous structures, which can be called disorganised structures [16]. The critical component in materials is the amorphous phase, which has been accepted as the essential component to undergo geopolymerisation. Despite the high silica and alumina content of the precursor, the rate of dissolution is determined by its structure. The greater the amorphous silica and alumina content, the greater the reactivity with alkaline solutions to undergo the dissolution process [1,17].

In this paper, the clay reactivity is studied, which can be improved by amorphization through de-hydroxylation. Metakaolin is the most famous raw material to be used in the manufacturing of geopolymers due to its high intensity of amorphous structure [18]. The presence of this glass phase content can be identified by the technique of X-ray diffraction, which determines the degree of crystallinity of the precursor [19]. Hence, it can be said that the amorphous silica and alumina content are the primary elements for the effective geopolymersation process instead of the high amounts of silica and alumina content if they are difficult to leach out by hydroxide ions.

Morphology

The finer the particles of a precursor, the greater the contact surface area of the particles with an alkaline medium and the higher the reactivity with an alkaline solution. The application of nanomaterials, which have high fineness properties, contributes to high reactivity with high rates of dissolution that results in high compressive strength. However, there is a limit to the amount added that will cause agglomeration that affects the homogeneous properties of the geopolymer [1]. Furthermore, the high surface area of particles will result in high compressive strength if compared to the low surface area of particles in raw materials [20]. The morphology can be identified by the technique of Scanning Electron Microscope-Energy (SEM) as shown in Fig. 2

Composite Materials: SEAJCCM2024
Materials Research Proceedings 56 (2025) 21-29

Materials Research Forum LLC
https://doi.org/10.21741/9781644903636-3

[21]. In short, enough fine precursors with high surface area led to high reactivity with the alkaline medium and contribute to a compact and denser geopolymer matrix.

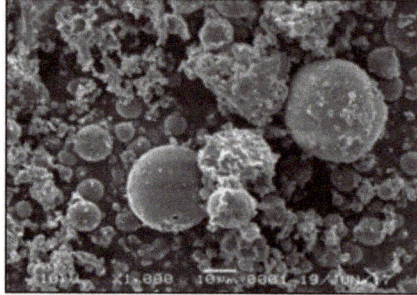

Fig. 2. Micro-structure of the fly ash particle.

Mineralogy

The composition of minerals in clay is critical because it may contain unreactive minerals that have a high degree of crystallisation and are difficult to react with an alkaline medium to form the desired properties of geopolymers. There are some minerals that should be noticed that possess high reactivity properties. For example, muscovite is a clayey mineral with high reactivity [22]. Furthermore, halloysite, which is present in kaolin, contributes to a high rate of geopolymerisation due to its tubular structure, which allows for the dissolution to occur inside and outside of the tube [23]. Besides, the quartz present in calcinated kaolin has high reactivity, which performed well in the geopolymerisation process and resulted in high mechanical properties [24]. The presence of these minerals can be identified by the technique of X-ray diffraction (XRD) [21], and an example of the result of the XRD is shown in Fig. 3. In short, the presence of different minerals in different precursors is one of the indicators of the reactivity of the precursor.

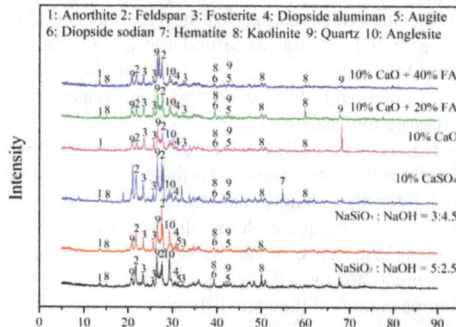

Fig. 3. XRD patterns of the solidified geopolymer blocks [25].

Ways for Improving Reactivity

Thermal, chemical, and mechanical are the methods that can enhance the reactivity of the precursor with the stabilizers. They can make the low-reactivity precursor highly reactive, which results in high strength and durability of the final product.

Thermal Treatment

The low reactivity of the precursor can be improved by undergoing calcination, which is treated at a high temperature within the range. The thermal treatment comprises dehydration, which is the

Composite Materials: SEAJCCM2024
Materials Research Proceedings 56 (2025) 21-29

Materials Research Forum LLC
https://doi.org/10.21741/9781644903636-3

continuous loss of interlayer water, and dehydroxylation, which is the discontinuous loss of structural water that results in a large reduction in the basal distance as well as interlayer gap buckling. There are also significant structural changes that occur as the bonding coordination numbers of Al atoms in the octahedral sheet degrade, making them more reactive. The stretched nature of the alumina layers is responsible for this material's strong reactivity, particularly in geopolymeric gel-forming conditions. It can be identified by the presence of disordered Al atom coordination [23].

For example, clay with kaolin is typically calcined between 600 and 900 degrees Celsius, resulting in metakaolin with a very reactive and disordered structure that has a higher ability to react with alkaline activators [26]. It can be concluded that the disordered structure of the precursor could be formed during calcination, which would increase its reactivity. It can be concluded that the disordered structure of the precursor could be formed by calcination, which increases its reactivity.

Chemical Treatment

The usual blending method involves mixing the basic material with a chemical addition. This approach modifies the source material's bulk chemical composition, especially in terms of the amounts of SiO_2 and Al_2O_3, resulting in a change in its geopolymeric reactivity. When $Al(OH)_3$ was added to a slag geopolymer system mixture, it increased Si-Al replacement and the formation of amorphous gels [27]. Furthermore, the extra N-A-S-H phases are formed, and there is a high rate of dissolution of Si and Si–Al phases with the addition of nanoparticles [1]. The addition of reactive substances such as slag can increase the less reactive geopolymer system [28]. Furthermore, the addition of metakaolin increases the amount of amorphous SiO_2 and Al_2O_3 [29]. Both of these additions can shorten the settling time and increase the mechanical properties of the geopolymer [30,31]. The addition of calcined kaolin clay increased the rate of geopolymer production and the compressive strength of the less reactive geopolymer source materials. In short, the blending method can be an alternative method for increasing the reactivity of the precursor that is less polluting and simpler than thermal treatment by mixing the addition of chemical or high-reactivity material.

Besides this blending technique, the alkaline fusion of source materials can also promote geopolymeric reactivity by calcining the raw material and sodium hydroxide mixture at a temperature greater than the melting point of NaOH [7]. It aided in the dissolution of Si and Al species from less reactive fly ash materials, increasing their reactivity [32]. It alters the raw material's mineralogical constitution and causes the production of an amorphous phase [33]. In addition, the addition of metakaolin increased the amount of reactive phase synthesized, resulting in the dissolution of more silicon and alumina species [34]. Overall, it is a complex process, but the reactivity of the precursors could be effectively enhanced, which has changed the mineralogical composition.

Mechanical Treatment

Mechanical activation is a method that improves the reactivity of a solid by applying mechanical energy without affecting its chemical composition [35]. It will increase the reactivity by causing crystal disordering and the formation of deficiencies, which lower the activation energy barrier for the process [36]. Overall, it will mostly affect particle size reduction, which causes alterations in physical properties [37].

The general method of this mechanical treatment is grinding. The grinding process used to decrease the crystallinity degree of the kaolin results in high reactivity with alkaline solutions, is a less polluting method, and is preferable to thermal treatment, which emits high levels of carbon dioxide [38]. Moreover, the reactivity of the natural clay can be enhanced by the mechanochemical activation method, which results in a high degree of disordered structure, a small particle size, and a high surface area of the particle, as shown in Fig. 4 [39,40,41]. Furthermore, kaolin that has undergone mechanochemical processing has a high rate of dissolution and an increased compressive strength [42]. In short, it is another method of increasing the reactivity of the precursor without altering the chemical constitution but altering the physical properties by decreasing the particle size, which results in a high rate of dissolution.

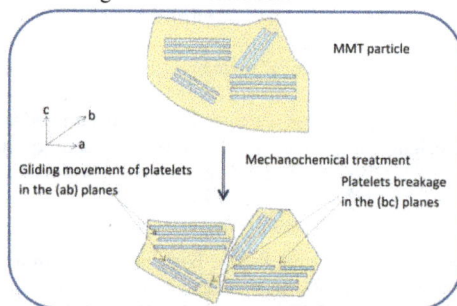

Fig. 4. Schematic illustration of the reduction in the size of clay particles after grinding [44].

Conclusion

In short, the properties of the precursor, such as its chemical composition, glass phase content, morphology, and mineralogy, should be investigated because they act as the main component part of the geopolymerisation process. It indicates the final structure of the geopolymer matrix and the properties in terms of mechanical and durability. However, if low reactivity of the precursor is encountered, there are three treatments, namely thermal, chemical, and mechanical, that could be applied to alter it to become highly reactive with the alkaline medium. Hence, it should be noted for manufacturing standard unfired clay brick in the construction field.

Acknowledgements

This research was funded by a grant from Ministry of Higher Education of Malaysia (FRGS Grant R.J130000.7824.4X172).

References

[1] M. Sumesh, U.J. Alengaram, M.Z. Jumaat, K.H. Mo, M.F. Alnahhal, Incorporation of nano-materials in cement composite and geopolymer based paste and mortar A review, Construction and Building Materials 148 (2017) 62-84. https://doi.org/10.1016/j.conbuildmat.2017.04.206

[2] Mucsi, Gábor, M. Ambrus. MultiScience - XXXI (2018). https://doi.org/10.26649/musci.2017.008

[3] S. Iftikhar, K. Rashid, E.U. Haq, I. Zafar, F.K. Alqahtani, M.I. Khan, Synthesis and characterization of sustainable geopolymer green clay bricks: An alternative to burnt clay brick, Construction and Building Materials 259 (2020) 119659. https://doi.org/10.1016/j.conbuildmat.2020.119659

[4] D. Muheise-Araalia, S. Pavia, Properties of unfired, illitic-clay bricks for sustainable construction, Construction and Building Materials 268 (2021) 121118. https://doi.org/10.1016/j.conbuildmat.2020.121118

[5] M. Zhang, H. Guo, T. El-Korchi, G. Zhang, M. Tao, Experimental feasibility study of geopolymer as the next-generation soil stabilizer, Construction and building materials 47 (2013) 1468-1478. https://doi.org/10.1016/j.conbuildmat.2013.06.017

Composite Materials: SEAJCCM2024
Materials Research Proceedings 56 (2025) 21-29

Materials Research Forum LLC
https://doi.org/10.21741/9781644903636-3

[6] T. Ye, J. Xiao, Z. Duan, S. Li, Geopolymers made of recycled brick and concrete powder – A critical review, Construction and Building Materials 330 (2022) 127232. https://doi.org/10.1016/j.conbuildmat.2022.127232

[7] L.N. Tchadjie, S.O. Ekolu, Enhancing the reactivity of aluminosilicate materials toward geopolymer synthesis, Journal of materials science 53 (2018) 4709-4733. https://doi.org/10.1007/s10853-017-1907-7

[8] S. Luhar, I. Luhar, D. Nicolaides, R. Gupta, Durability performance evaluation of rubberized geopolymer concrete, Sustainability 13 (2021) 5969. https://doi.org/10.3390/su13115969

[9] A. Nikolov, I. Rostovsky, H. Nugteren, Geopolymer materials based on natural zeolite. Case Studies in Construction Materials 6 (2017) 198-205. https://doi.org/10.1016/j.cscm.2017.03.001

[10] R. Mohamed, R. Abd Razak, M.M.A.B. Abdullah, R.K. Shuib, J. Chaiprapa, Geopolymerization of class C fly ash: reaction kinetics, microstructure properties and compressive strength of early age, Journal of Non-Crystalline Solids 553 (2021) 120519. https://doi.org/10.1016/j.jnoncrysol.2020.120519

[11] L. Hou, J. Li, Z.Y. Lu, Effect of Na/Al on formation, structures and properties of metakaolin based Na-geopolymer, Construction and Building Materials 226 (2019) 250-258. https://doi.org/10.1016/j.conbuildmat.2019.07.171

[12] C.R. Kaze, S.B.K. Jiofack, Ö. Cengiz, T.S. Alomayri, A. Adesina, H. Rahier, Reactivity and mechanical performance of geopolymer binders from metakaolin/meta-halloysite blends, Construction and Building Materials 336 (2022) 127546. https://doi.org/10.1016/j.conbuildmat.2022.127546

[13] M.I. Morsy, K.A. Alakeel, A.E. Ahmed, A.M. Abbas, A.I. Omara, N.R. Abdelsalam, H.H. Emaish, Recycling rice straw ash to produce low thermal conductivity and moisture-resistant geopolymer adobe bricks, Saudi Journal of Biological Sciences 29 (2022) 3759-3771.

[14] U. Ghani, S. Hussain, M. Imtiaz, S.A. Khan, Role of calcination on geopolymerization of lateritic clay by alkali treatment, Journal of Saudi Chemical Society 25 (2021) 101198. https://doi.org/10.1016/j.jscs.2021.101198

[15] M. Ahmad, K. Rashid, Novel approach to synthesize clay-based geopolymer brick: Optimizing molding pressure and precursors' proportioning, Construction and Building Materials 322 (2022) 126472. https://doi.org/10.1016/j.conbuildmat.2022.126472

[16] K.Z. Farhan, M.A.M. Johari, R. Demirboğa, Assessment of important parameters involved in the synthesis of geopolymer composites: A review, Construction and Building Materials 264 (2020) 120276. https://doi.org/10.1016/j.conbuildmat.2020.120276

[17] C.R. Kaze, A. Adesina, G.L. Lecomte-Nana, H. Assaedi, T. Alomayri, E. Kamseu, U.C. Melo, Physico-mechanical and microstructural properties of geopolymer binders synthesized with metakaolin and meta-halloysite as precursors, Cleaner Materials 4 (2022) 100070. https://doi.org/10.1016/j.clema.2022.100070

[18] J. Bensted, P. Barnes, Structure and Performance of Cements, New York: Spon Press, 2002. https://doi.org/10.1201/9781482295016

[19] N. Sedira, J. Castro-Gomes, M. Magrinho, Red clay brick and tungsten mining waste-based alkali-activated binder: Microstructural and mechanical properties, Construction and Building Materials 190 (2018) 1034-1048. https://doi.org/10.1016/j.conbuildmat.2018.09.153

[20] H.K. Tchakoute, A. Elimbi, E. Yanne, C.N. Djangang, Utilization of volcanic ashes for the production of geopolymers cured at ambient temperature, Cement and Concrete Composites 38 (2013) 75-81. https://doi.org/10.1016/j.cemconcomp.2013.03.010

[21] K. Bouguermouh, N. Bouzidi, L. Mahtout, L. Pérez-Villarejo, M.L. Martínez-Cartas, Effect of acid attack on microstructure and composition of metakaolin-based geopolymers: The role of alkaline activator, Journal of Non-Crystalline Solids 463 (2017) 128-137. https://doi.org/10.1016/j.jnoncrysol.2017.03.011

[22] H.K. Tchakouté, S. Kong, J.N.Y. Djobo, L.N. Tchadjié, D. Njopwouo, A comparative study of two methods to produce geopolymer composites from volcanic scoria and the role of structural water contained in the volcanic scoria on its reactivity, Ceramics International 41 (2015) 12568-12577. https://doi.org/10.1016/j.ceramint.2015.06.073

[23] A.Z. Khalifa, Ö. Cizer, Y. Pontikes, A. Heath, P. Patureau, S.A. Bernal, A.T. Marsh, Advances in alkali-activation of clay minerals, Cement and Concrete Research 132 (2020) 106050. https://doi.org/10.1016/j.cemconres.2020.106050

[24] H.K. Tchakoute, C.H. Rüscher, J.Y. Djobo, B.B.D. Kenne, D. Njopwouo, Influence of gibbsite and quartz in kaolin on the properties of metakaolin-based geopolymer cements, Applied Clay Science 107 (2015) 188-194. https://doi.org/10.1016/j.clay.2015.01.023

[25] D. Song, T. Huang, Q. Fang, A. Liu, Y. F. Gu, Y.Q. Liu, L.F. Liu, S.W. Zhang, Feasibility exploration on the geopolymerization activation of volcanic tuff, parametrical optimization, and reaction mechanisms, Journal of Materials Research and Technology 11 (2021) 618-632. https://doi.org/10.1016/j.jmrt.2021.01.029

[26] T. Kovářík, P. Bělský, P. Novotný, J. Říha, J. Savková, R. Medlín, D. Rieger, P. Holba, Structural and physical changes of re-calcined metakaolin regarding its reactivity, Construction and Building Materials 80 (2015) 98-104. https://doi.org/10.1016/j.conbuildmat.2014.12.062

[27] M.O. Yusuf, M.A.M. Johari, Z.A. Ahmad, M. Maslehuddin, Effects of addition of Al (OH)3 on the strength of alkaline activated ground blast furnace slag-ultrafine palm oil fuel ash (AAGU) based binder, Construction and Building Materials 50 (2014) 361-367. https://doi.org/10.1016/j.conbuildmat.2013.09.054

[28] S. Kumar, R. Kumar, S.P. Mehrotra, Influence of granulated blast furnace slag on the reaction, structure and properties of fly ash based geopolymer, Journal of materials science 45 (2010) 607-615. https://doi.org/10.1007/s10853-009-3934-5

[29] R.A. Robayo-Salazar, R.M. De Gutiérrez, F. Puertas, Effect of metakaolin on natural volcanic pozzolan-based geopolymer cement, Applied Clay Science 132 (2016) 491-497. https://doi.org/10.1016/j.clay.2016.07.020

[30] S. Saha, C. Rajasekaran, Enhancement of the properties of fly ash based geopolymer paste by incorporating ground granulated blast furnace slag, Construction and Building Materials 146 (2017) 615-620. https://doi.org/10.1016/j.conbuildmat.2017.04.139

[31] J.Y. Djobo, L.N. Tchadjié, H.K. Tchakoute, B.B.D. Kenne, A. Elimbi, D. Njopwouo, Synthesis of geopolymer composites from a mixture of volcanic scoria and metakaolin, Journal of Asian Ceramic Societies 2 (2014) 387-398. https://doi.org/10.1016/j.jascer.2014.08.003

[32] H. Xu, Q. Li, L. Shen, M. Zhang, J. Zhai, Low-reactive circulating fluidized bed combustion (CFBC) fly ashes as source material for geopolymer synthesis, Waste Management 30 (2010) 57-62. https://doi.org/10.1016/j.wasman.2009.09.014

[33] H.K. Tchakoute, A. Elimbi, B.D. Kenne, J.A. Mbey, D. Njopwouo, Synthesis of geopolymers from volcanic ash via the alkaline fusion method: Effect of Al2O3/Na2O molar ratio of soda–volcanic ash, Ceramics International 39 (2013) 269-276. https://doi.org/10.1016/j.ceramint.2012.06.021

[34] H.T. Kouamo, A. Elimbi, J.A. Mbey, C.N. Sabouang, D. Njopwouo, The effect of adding alumina-oxide to metakaolin and volcanic ash on geopolymer products: A comparative study, Construction and Building Materials 35 (2012) 960-969. https://doi.org/10.1016/j.conbuildmat.2012.04.023

[35] P. Baláž, M. Achimovičová, M. Baláž, P. Billik, Z. Cherkezova-Zheleva, J.M. Criado, F. Delogu, K. Wieczorek-Ciurowa, Hallmarks of mechanochemistry: from nanoparticles to technology, Chemical Society Reviews 42 (2013) 7571-7637. 10.1039/c3cs35468g

[36] V.V. Boldyrev, K. Tkáčová, Mechanochemistry of solids: past, present, and prospects, Journal of materials synthesis and processing 8 (2000) 121-132. https://doi.org/10.1023/A:1011347706721

[37] G. Mucsi, Mechanical activation of power station fly ash by grinding–A review, Építőanyag 68 (2016) 56-61. https://doi.org/10.14382/epitoanyag-jsbcm.2016.10

[38] I. Tole, K. Habermehl-Cwirzen, A. Cwirzen, Mechanochemical activation of natural clay minerals: an alternative to produce sustainable cementitious binders–review, Mineralogy and Petrology 113 (2019) 449-462. https://doi.org/10.1007/s00710-019-00666-y

[39] P.J. Sanchez-Soto, A. Wiewiora, M.A. Avilés, A. Justo, L.A. Pérez-Maqueda, J.L. Pérez-Rodríguez, P. Bylina, Talc from Puebla de Lillo, Spain. II. Effect of dry grinding on particle size and shape, Applied Clay Science 12 (1997) 297-312. https://doi.org/10.1016/S0169-1317(97)00013-6

[40] I. Bekri-Abbes, E. Srasra, Effect of mechanochemical treatment on structure and electrical properties of montmorillonite, Journal of Alloys and Compounds 671 (2016) 34-42. https://doi.org/10.1016/j.jallcom.2016.02.048

[41] J. Ondruška, Š. Csáki, V. Trnovcova, I. Štubňa, F. Lukáč, J. Pokorný, L. Vozar, P. Dobroň, Influence of mechanical activation on DC conductivity of kaolin, Applied Clay Science 154 (2018) 36-42. https://doi.org/10.1016/j.clay.2017.12.038

[42] A. Souri, H. Kazemi-Kamyab, R. Snellings, R. Naghizadeh, F. Golestani-Fard, K. Scrivener, Pozzolanic activity of mechanochemically and thermally activated kaolins in cement, Cement and Concrete Research 77 (2015) 47-59. https://doi.org/10.1016/j.cemconres.2015.04.017

Composite Materials: SEAJCCM2024
Materials Research Proceedings 56 (2025) 30-42

Materials Research Forum LLC
https://doi.org/10.21741/9781644903636-4

Assessing CO_2 Emissions Potential Using Satellite Data: A Case Study of Malaysia's Highlands

Norhusna Mohamad NOR[1,a] * and Nur Syifa SALIM[1,b]

[1]Chemical Engineering Studies, Universiti Teknologi MARA, Cawangan Pulau Pinang, Permatang Pauh Campus, 13500 Pulau Pinang, Malaysia

[a]norhusna8711@uitm.edu.my, [b]syifasalim07@gmail.com

Keywords: CO_2 Emissions, Malaysia's Highlands, Remote Sensing, Satellite Database, Grey Relation Analysis (GRA)

Abstract. Environmental issues are a significant concern impacting the ecosystem, and analysing carbon dioxide (CO_2) emissions in Malaysia's highlands is crucial. Most spatial regions in the highlands in Malaysia were significantly affected by the excessive release of CO_2 due to various factors significantly contributed to by dynamic industrial activities. This phenomenon may also be affected by the changes in other external variables in the highlands, such as air temperature and atmospheric moisture. The Grey Relation Analysis (GRA) approach is proposed in this study to investigate the correlation between CO_2 emissions and multiple variables. The findings utilizing the GRA method are important in understanding other variables that contributed to the increase in CO_2 emissions. Hence, a suitable strategy can be suggested to control the variables and substantially lower CO_2 emissions. To accomplish this, Giovanni, a satellite database remote sensing, has been selected and used to gather all related data within Cameron Highland, Genting Highland, Kundasang, and Kelabit Highland. The comparison between the multiple variables shows that the proposed method has higher accuracy, which presents that the correlation between atmospheric moisture and CO_2 emissions has significant outcomes in proving the variables that have been affected by the CO_2 emissions. The analysed results via the GRA method show that the most affected variable due to the CO_2 emissions is atmospheric moisture, where Cameron Highland states the higher grey relational coefficient ($\gamma = 2.8125$) compared to the other associated factors within the other highlands.

Introduction

Malaysia's highlands are among the destinations where you can relax after a long day at work. Most of the highlands in Malaysia were grown throughout the fragile and vulnerable forested area, which consists of low temperatures and high rainfall amounts. There are many highlands in Malaysia nowadays undergoing urbanisation in terms of housing, plantation, recreational, and industry activities, such as Genting Highlands, Fraser's Hill, Kundasang and Cameron Highlands. Most of the highlands undergo advanced development and rapid transformation to achieve an urban standard of living in Malaysia. Hence, most of the forest fronting damage caused Malaysia's highlands to be one of the locations on offer to unwind after a long day at work. The three most well-known highlands that are consistently a draw for a tranquil getaway are Fraser's Hill, Cameron Highlands, and Genting Highlands. Malaysia's Highlands is one of the destinations on offer to unwind after a long day at work by selective logging, tea plantation for tourism purposes, farming, and so on [1]. From those issues, the atmosphere is consequently exposed to the excessive release of greenhouse gases, especially CO_2 emissions due to deforestation. This indirectly will lead to a rise in temperature and adverse effects on the environment and human health. Global warming, climate change, and other environmental problems are linked to excess human activities contributing to the depletion of the stratospheric ozone layer [2]. Tan and Loh [3] evaluated the

Content from this work may be used under the terms of the Creative Commons Attribution 3.0 license. Any further distribution of this work must maintain attribution to the author(s) and the title of the work, journal citation and DOI. Published under license by Materials Research Forum LLC.

Composite Materials: SEAJCCM2024   Materials Research Forum LLC
Materials Research Proceedings 56 (2025) 30-42   https://doi.org/10.21741/9781644903636-4

CO_2 emissions in Cameron Highlands using the Downscaling approach, where it was found that Cameron Highlands experienced significant climate change projections from 1980 to 2069.

Compared to other greenhouse gases such as methane (CH_4) and nitrous oxide (NO_x), CO_2 is among the most abundantly emitted. This is because CO_2 has a longer lifetime than CH_4 and NO_x, which increases the average surface air temperature [3]. Generally, excessive human activities, including the manufacturing of pesticides, deforestation, synthetic fertilisers, changes in land use, and agricultural lime, are the main contributors to CO_2 emissions. Other significant variables that are also affecting the CO_2 emissions include urbanisation, which is occurring in most cities and specific regions, such as the highlands [4]. Most of the global energy demand will also create tremendous CO_2 emissions through the atmosphere. In 2015, The Ministry of Natural Resources and Environment reported the increment of CO_2 emissions in Malaysia's atmosphere, which were 73, 76 and 72% of the total greenhouse gas emissions in 2000, 2005, and 2011 respectively [5]. In 2018, with higher energy demand for electrical consumption and industrial sector purposes, the CO_2 emissions rose by about 1.7% and slowly rose continuously [6].

Numerous detection systems, such as satellite remote sensing for data collection on CO_2 emissions, were developed to address this issue. Satellite database remote sensing is one of the technologies utilised for image-based data collecting and analysis [7]. Various data types can be collected using satellite database remote sensing comprising air pressure, air temperature, CO_2, CO, CH_4 and so on [8]. All the satellite database remote sensing is a web-based tool that collects and visualizes data for all categories, including CO_2 emissions [9]. Recently, CO_2 emissions observations have mainly depended on the satellite database remote sensing, including Japan's Greenhouse Gases Observing Satellite (GOSAT), NASA's Orbiting Carbon Observatory – 2 (OCO-2), China's Carbon Satellite and Giovanni [12]. In all those satellite databases, remote sensing is the standard method used to collect CO_2 concentrations based on some regional regions. For instance, from Giovanni tools, data can be collected variously based on either the latitude or longitude of the map, scatter plots, and time series plots [9]. Data on air quality and CO_2 can be easily collected using the satellite database remote sensing. Table 1 shows the performance and comparison of available satellite database remote sensing in data acquisition of CO_2 emissions.

Table 1. Highlands' coordinates used in Giovanni via time series-area averaged.

Sensors	On board satellites	Orbital altitude (km)	Spatial resolution (km)	Spectral region (μm)	Observation made	References
GOSAT – TANSO	GOSAT	666	10.5	0.75 – 14.3	Nadir, flare, target	[10]
Giovanni	Terra	705	0.25	0.405 – 14.385	Nadir	[11]
SCIAMACHY	ENVISAT	799.8	30 × 60	0.24 – 2.38	Limb, Nadir	[12]
AIRS	Aqua	705.3	13.5	3.7 – 15.4	Nadir	[12]

From Table 1, it can be concluded that the Giovanni web-based tool has the highest spectral region in evaluating CO_2 emissions, which is between 0.405 and $14.385 \mu m$. This spectral region is the optimum range for simulations of global and regional scenes. This system is principally expected to predict and evaluate the global changes in the environment due to specific anthropogenic activity [13]. In Giovanni, the detecting instrument used was a thermal channel, which focused more on the temperature and moisture of the earth. Thus, Giovanni is the most suitable multifunction remote sensing that can determine the factors contributing to CO_2 emissions, such as air temperature changes and atmospheric moisture.

Composite Materials: SEAJCCM2024   Materials Research Forum LLC
Materials Research Proceedings 56 (2025) 30-42   https://doi.org/10.21741/9781644903636-4

Data analysis is one of the tools that can be used to study various factors that might influence the CO_2 emissions in Malaysia's highlands. The Grey Relation Analysis (GRA), which may be used for the prediction and evaluation of CO_2 emissions, is a correlation coefficient approach that is employed in determining the factors that contribute to the one selected parameter [14]. The GRA is also one of the qualitative and quantitative analyses used to determine the non-figurative relationship [15]. The CO_2 emission can be statistically analysed and correlated to their factors, such as the air temperature changes by assuming the data distribution as linear, exponential or logarithmic, and errors that are distributed [16]. The strong corresponded between the air temperature changes and the CO_2 emissions can be shown by the result obtained from the analysis method of GRA [17]. The main objective of this work is to investigate the significant variables affected by the release of CO2 emissions in the selected Malaysian highlands.

CO_2 Emissions

Day to day, CO_2 emissions are commonly and exaggeratedly being released due to inappropriate human activities, making the earth warmer due to global warming issues. There are some different types of activities that, in the end, can consequently cause CO_2 emissions. It was reported that all the usage of coal, oil, and natural gas could undoubtedly cause an increase in CO_2 emissions, which was about 1.4% globally in 2017 [21]. Electricity and heat production, manufacturing industries, construction, and transportation sectors can also be the main reasons contributing to CO_2 emissions, as shown in Fig. 1. Electricity and heat production contribute more than half a million tons of CO_2 emission equivalent, about 103.1 million tons. Electricity is the most urgent energy demand, and the rising pace has been faster.

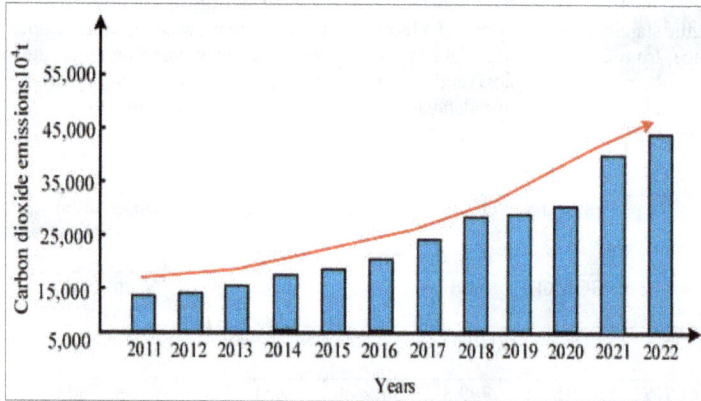

Fig. 1. Carbon Dioxide Emissions by Sectors [18].

Most of the global energy demand will also create tremendous CO_2 emissions through the atmosphere. Besides electricity and heat production, some industrial sectors also need to use non-renewable energy for their processes, such as gas, oil, coal, and nuclear. Thus, it becomes more relatable for the more excellent production of CO_2. Based on Fig. 2, the study shows that the consumption of global energy demand rose in 2018, about twice the average rate of growth since 2010 [6]. Global energy demand has increased due to the higher electricity demand caused by urbanisation, which has happened throughout the years. As a result of higher energy demand for electrical consumption and industrial sector purposes, the CO_2 emissions rose by about 1.7% in 2018 and slowly rose continuously without the realisation that people surrounding [6].

Composite Materials: SEAJCCM2024
Materials Research Proceedings 56 (2025) 30-42

Materials Research Forum LLC
https://doi.org/10.21741/9781644903636-4

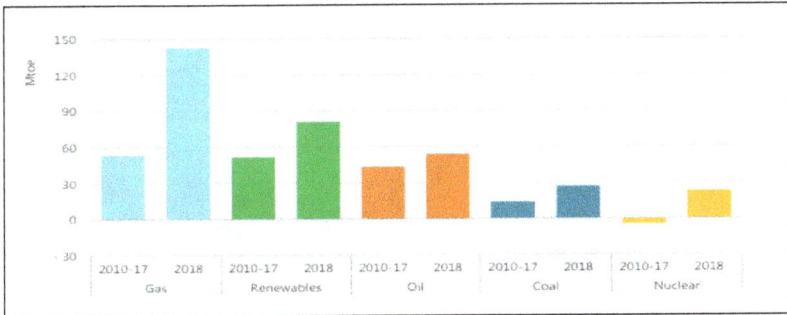

Fig. 2. Average annual global primary energy demand growth by fuel, 2010-18 [19].

CO_2 Emissions in Malaysia's Highlands

Highlands in Malaysia is one of the offering places to release the mind after loads of work. Cameron Highlands, Genting Highlands, and Fraser's Hill are the most popular highlands that will always be attractions for a relaxing destination. Most of the highlands were grown throughout the fragile and vulnerable forested area, which consists of low temperatures and high rainfall amount [19]. As highlands undergo development, more activities on urbanization take place in terms of agricultural and sustainable tourism [19].

Cameron Highlands and Genting Highlands are some of the highlands located in Peninsular Malaysia. They are located in the small district of Pahang and the Main Range (Banjaran Titiwangsa) within the coordinates of 4°20'N – 4°37'N, 101°20' - 101°36' E for Cameron Highland and 3.4239° N, 101.7927 E for Genting Highland respectively [20]. Fig. 3 shows the exact location of the Cameron Highlands, Genting Highlands, Kelabit Highlands and Kundasang in the map view. Agricultural and transportation sectors can be claimed as highly demand sectors in today's population, and all the highlands stated are some of the highlands that are well-known for the agricultural and transportation activities such as tea leaf, flowers, fruits and varieties of vegetable plantation [20]. Most of the previous studies focused on agricultural and transportation sectors because they are one of the reasons linked to climate change and consequently will cause the temperature to rise, and carbon fertilization affects that primarily due to the CO_2 emissions [21]. Besides, urban development also contributes a lot to the overabundance of environmental problems that often happen [22]. Urbanization happens, including the development of housing areas and plantation areas, which require the forest to be encroached [22]. From forest encroachment, the crisis of deforestation becomes wild and consequently will lead to higher CO_2 emissions.

Another highland that can be found in Borneo, which is undergoing thriving development, is Kundasang and Kelabit Highland. Kundasang is located in Sabah Highlands on the southeast side of Gunung Kinabalu, while Kelabit Highland is located northeast of Sarawak [23]. It is the most developed area since it is renowned as a tourist attraction, and transportation is one of the main issues relating to CO_2 emissions because the increase in urban areas in Sabah will lead to many vehicles on their roadway. Almost 90% of CO_2 emissions were from motor vehicles on the roadway of Sabah, especially in urban areas such as Kundasang [24].

Composite Materials: SEAJCCM2024   Materials Research Forum LLC
Materials Research Proceedings 56 (2025) 30-42   https://doi.org/10.21741/9781644903636-4

Fig. 3. Location of Highlands in Peninsular Malaysia and Borneo.

Satellite Database in CO_2 Emissions

In this research, the primary key to combating environmental issues is collecting data on CO_2 emissions. Two types can be used to determine and collect data on CO_2 emissions worldwide by ground-based and satellite-based methods, respectively. Compared to ground-based methods, satellite-based methods can be eminently known as convenient methods for today's generation because satellite databases make coverage more comprehensive. For the ground-based method, the forecasting method to forecast CO_2 emissions has been used in most previous studies. For instance, an Artificial Neural Network (ANN) has been set up in Serbia to estimate CO_2 emissions by considering the issues of CO_2 emissions, such as mitigation of CO_2 emissions [25]. The satellite-based method can easily collect data from satellite database remote sensing such as GOSAT-TANSO, SCIAMACHY, AIRS, and Giovanni. All the satellite database remote sensing is a web-based tool that collects and visualizes data for all categories and CO_2 emissions included in the categories [26]. All the satellite database remote sensing performances were compared, including the onboard satellites' orbital altitude, spatial resolution, spectral region, detection instrument, and observation made.

Satellite database remote sensing is one of the tools that will be conveniently used to collect precise data on CO_2 emissions in Malaysia's highlands. This study will focus more on Giovanni in collecting and analyzing data on CO_2 emissions in Malaysia's highlands, such as Cameron Highland and Kundasang. Giovanni is the web-based tool used to visualize and analyses all the data regarding Earth science, such as CO_2, NO_X and CH_4 [14]. The tools available for monitoring land use change and the deforestation of forests suit the research study regarding CO_2 emissions in Malaysia's highlands. Visualizing the tools will create a live visual of the land covered, including forest areas in which deforestation detection will quickly occur. Table 2 compares common satellite database remote sensing used for specific purposes in collecting data.

Table 2 compares satellite databases commonly used in certain countries and their purpose. AAU Sat-4 is one of the standard satellite databases for remote sensing used in Denmark; it was launched from Aalborg University Cubesat 4 and is commonly used in earth observation, especially in the monitoring activities for large sea areas [29]. In Japan, the satellite database

Composite Materials: SEAJCCM2024   Materials Research Forum LLC
Materials Research Proceedings 56 (2025) 30-42   https://doi.org/10.21741/9781644903636-4

remote sensing used commonly is GOSAT-TANSO, which also covers earth observation, focusing more on greenhouse gas parameters such as CO_2 and CH_4 [30]. For Malaysia, the common satellite database remote sensing that has been used was RazakSat, and the goal of having the satellite is observing the optical imaging by increasing the manufacturing competence and spacecraft design to achieve international development [31]. Giovanni satellite database remote sensing is a multinational satellite database remote sensing that can be used for earth science. It can be used to determine CO_2 emissions in Malaysia or a foreign country. It is also one of the standard satellite databases for remote sensing being used internationally.

Table 2. Comparison of Satellite Database Remote Sensing.

Country	Satellite Database Remote Sensing	Purpose	Reference
Denmark	AAU Sat – 4	Earth Observation	[27]
Japan	GOSAT–TANSO	Earth Observation	[27]
Malaysia	RazakSat	Optical Imaging	[28]
Multinational	Giovanni	Earth Science	[28]

Models Used for CO_2 Assessment

In this study of CO_2 emissions in Malaysia's highlands, all the factors can be determined by performing the data collection of the CO_2 emissions and continuously further analysing the data. In analyzing collected data, statistical analysis takes place, which is the correlation coefficient. For this study, the Grey Correlation Analysis method will be applied for the statistical analysis of the data. It is an accurate quantitative method showing the relationship between the different data sequences and the significantly different statistics [32]. At the end of this study, the grey relative correlation coefficient between CO_2 emissions and output value was determined using software calculation tools such as Excel and Analyse-it. Table 3 compares analysis results in highlands between different analysis methods in various study areas.

Table 3. Model Analysis Tools used in specific Study Areas.

Study Area	Model Analysis Tools	Variables	References
Malaysia (KL)	Multiple Linear Regression	Agriculture and CO_2 emissions	[33]
China (Hefei)	Stochastic Impact by Regression on Population, Affluence, and Technology (STIRPAT)	Population and Gross Domestic Product (GDP) and CO_2 emissions	[34]
Taiwan	Correlation Coefficient Method (GRA)	Temperature and CO_2 emissions	[35]
Malaysia (Cameron Highland)	Pearson Correlation Coefficient	Temperature and CO_2 emissions	[36]

From Table 3, different model analysis tools were used in various study areas depending on the parameters and relation of the data. For the CO_2 emissions data analysis, a standard method used in most countries to determine CO_2 is the correlation coefficient method, such as the Grey Analysis Method and Pearson Correlation Coefficient. All the standard methods, such as the correlation coefficient method, can be interpreted as:

$$\xi i0 = \frac{\min i\ (\min k)\ |x0\ (k) - xi(k)| + \rho\ \max i\ maxk\ |x0(k) - xi(k)|}{|x0(k) - xi(k)| + \rho\ \max i\ maxk\ |x0(k) - xi(k)|}$$

Eq. 1

where $\xi i0$ = Relational coefficient, min i min k |x0(k)−xi(k)| [max i max k |x0(k)−xi(k)|] = The absolute difference between two sequences. The correlation coefficient method shows the relationship between the variables that determine the study's quantitative analysis.

Methods to Assess CO_2 Emissions in Malaysia's Highlands
The overall process flow of this research work is illustrated in Fig. 4. A detailed explanation of the methods is enclosed in the following subsections.

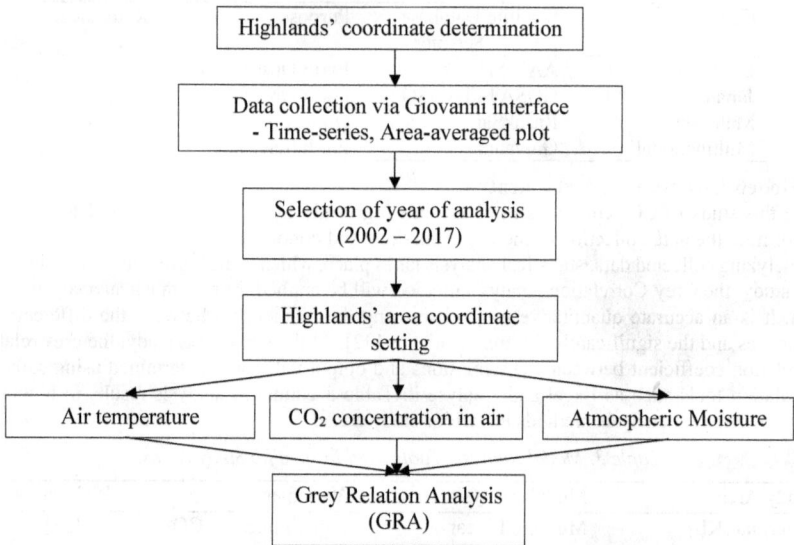

Fig. 4. Process flow chart of overall remote sensing data acquisition via Giovanni interface.

Highlands Coordinate Determination
The coordinates of all selected highlands were determined before accessing the required data from the Giovanni satellite database. The latitude and longitude information of the highlands location were identified using the Global Positioning System (GPS) and Google Maps. Table 4 outlines the coordinates for all selected highlands.

Table 4. Highlands' coordinates used in Giovanni via time series-area averaged.

Highlands	Highlands' Coordinates
Cameron Highland	101.6302E, 4.6351N, 101.9158E, 4.8823N
Genting Highland	101.7566E, 3.3115N, 101.8335E, 3.5422N
Kundasang	116.4673E, 5.8184N, 116.687E, 6.0162N
Kelabit Highland	114.3579E, 4.0915N, 114.4239E, 4.2233N

Giovanni Data Collection
The CO_2 emissions data was obtained using Giovanni web-based tools depending on the preference results. The data range selection was made from 2002 to 2017. The selected highlands' region/area based on the bounding box option in the Giovanni interface were identified based on

Materials Research Forum LLC
https://doi.org/10.21741/9781644903636-4

the longitude and latitude coordinates of the highlands, as shown in Fig. 5. The available variable corresponding to the CO₂ emissions data is the plotted data of CO₂ and mole fraction in free troposphere data in part per million (ppm). The CO₂ emissions values were obtained using time series, area-averaged plot, and the measurement of CO2, mole fraction (ppm) in the free troposphere, with a spatial resolution of 2 x 2.5 deg. [AIRS AIRX3C2M v005]. Other variables, such as air temperature and atmospheric moisture, were also obtained.

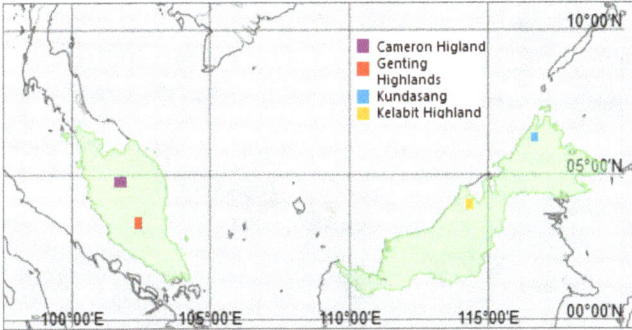

Fig. 5. The selected region/area map is based on the bounding box in the Giovanni interface

Data Analysis - Grey Relation Analysis

The correlation of CO₂ emissions with other variables was evaluated using Grey Relation Analysis (GRA). Eq. 1 and 2 were used to determine the maximum and minimum difference between the primary and associated factors [7]. For the grey relational coefficient (γ) The difference between the primary and associated factors in period k was calculated using Eq. 3. To determine the factors most affected by CO₂ emission, the grey relational degree was calculated using Eq. 4.

$$\Delta \min = min|x_0(k) - x_j(k)| \tag{1}$$

$$\Delta \max = max|x_0(k) - x_j(k)| \tag{2}$$

$$\gamma[x_0(k), x_i(k)] = \frac{\Delta \min + \zeta \Delta \max}{\Delta_{0i}(k) + \zeta \Delta \max} \tag{3}$$

$$\gamma(x_0, x_i) = \frac{1}{m} \sum_{k=1}^{m} \gamma[x_0(k), x_i(k)] \tag{4}$$

Where $x_0(k)$ is primary factor (CO₂ concentration), $x_j(k)$ is the associated factors (Air temperature, Atmospheric Moisture), ζ is the identification coefficient (0.5), and $\Delta_{0i}(k) | x_0(k) - x_i(k) |$

Assessment of CO₂ Emissions in Malaysia's Highlands Utilizing Giovanni

Fig. 6 shows the time series and area – average of the CO₂ mole fraction in the free troposphere. Table 3 tabulates the minimum, maximum, and the % increase of CO₂ emission in the selected highlands. In this case, the spatial resolution used was 2 x 2.5 deg. This is important as it influences how sharply human naked eyes can see the image. From the results obtained, the area–averaged CO₂ concentration for all the highlands is most likely to surpass the threshold values of the pre-industrial period, which was around 280 ppm [16]. Fig. 6 shows that within 15 years, a significant increment of CO₂ was emitted. Data tabulated in Table 5 validated that the range of CO₂ emissions increased by about 10 – 11% from 2002 to 2017. The highest area–averaged CO₂ concentration observed is in Kundasang, with CO₂ emissions of 412.8 ppm, recorded in November 2016. Various

Composite Materials: SEAJCCM2024   Materials Research Forum LLC
Materials Research Proceedings 56 (2025) 30-42   https://doi.org/10.21741/9781644903636-4

factors have contributed to the rise of the observed CO_2 emissions, among the factors resulting from the development activities in the highlands, such as deforestation, farming, transportation, and housing activities. Some of these activities may be due to illegal logging, mainly in Borneo and some of the highlands in Peninsular Malaysia. Borneo lost an average of 850,000 hectares of forest every year between 1985 and 2005 [37]. It was reported that all the usage of coal, oil, and natural gas could undoubtedly cause an increase in CO_2 emissions, which was about 1.4 globally in 2017 [38].

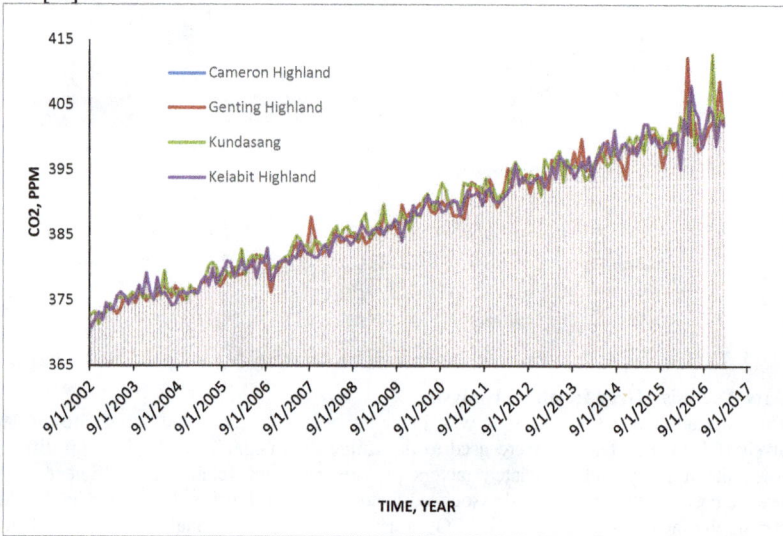

Fig. 6. Time series, area - Averaged of CO_2 emissions in Cameron Highland, Genting Highland, Kundasang, and Kelabit Highland from 2002 to 2017.

Table 5. Minimum, maximum, and % increase of CO2 emissions in the selected highlands.

Highlands	Min. CO_2 emissions, ppm	Max. CO_2 emissions, ppm	% CO_2 increase
Cameron Highland	371.000	412.323	11.1383
Genting Highland	371.000	412.323	11.1383
Kundasang	371.340	412.892	11.1899
Kelabit Highland	370.839	408.039	10.0314

Agriculture and transportation are two industries in high demand, and all the selected highlands are well-known for their farming and transportation activities, such as tea leaves, flowers, fruit, and vegetable plantations [15]. Most of the earlier research concentrated on the agricultural and transportation industries because they are two key contributors to climate change, which will lead to CO_2 emissions. Additionally, the excessive number of frequent environmental issues is greatly influenced by urban expansion [16]. Urbanisation includes the construction of housing developments and plantation lands, both of which encroach on the forest. Due to the forest encroachment, deforestation has gotten out of control and is now driving up CO_2 emissions. Instead of CO_2 emissions, these external factors contribute to other related variables. Among the variables that can be observed are the rise of air temperature and atmospheric moisture of the

Composite Materials: SEAJCCM2024   Materials Research Forum LLC
Materials Research Proceedings 56 (2025) 30-42   https://doi.org/10.21741/9781644903636-4

highlands. The findings were also supported by data analysed by Bekhet and Othman [39], where, based on ecological modernisation and augmented Cobb-Douglas production theories, the unidirectional causality from urbanisation to CO_2 emissions in the short run was found at a 1% significance level. In the long run, the bidirectional causality between CO_2 emissions and urbanisation is at a 5% level of significance.

Analysis of Associated Variables using GRA

Deng Julong introduced the Grey theory, a system science theory, for the first time in 1982 [40]. The theory encompassed significant advancements in the study of uncertainty systems, proving the connection between two variables with an ambiguous relationship. The principle of the Grey theory utilises the geometrical resemblance of the time series of the two variables—the more significant the correlation between the two variables, the more significant the variable. The correlation strength between several variables and the same reference sequence was used to identify dominant factors/variables.

Associated variables which related to the CO_2 emissions were evaluated using the GRA approach. Table 6 tabulates the Grey Relational coefficient (γ) values obtained from the correlation of the main parameter, CO_2 concentration, and another two (2) associated variables, which are air temperature and atmospheric moisture for the selected Malaysia's highlands. Overall, the values acquired indicate that the correlation between the CO_2 concentration and atmospheric moisture was higher γ compared to the air temperature. For Cameron Highland, the associated variable affected critically by the CO_2 emissions is atmospheric moisture, in which the γ for this associated variable is higher than air temperature. Cameron Highland has a higher γ value than Genting Highlands, Kundasang, and Kelabit Highland, with $\gamma = 2.8125$. On the contrary, the atmospheric moisture of Genting Highlands and Kelabit Highland is less affected by CO_2 emissions compared to Cameron Highland and Kundasang.

Table 6. Grey Relational coefficient (γ) of CO_2 concentration with associated variables.

Highlands	γ of air temperature (K)	γ of atmospheric moisture (kg/kg)
Cameron Highland	0.9880	2.8125
Genting Highlands	0.9254	1.5345
Kundasang	0.9941	2.0598
Kelabit Highland	1.0153	1.6699

The results show that the atmospheric moisture of Cameron Highland was notably affected by the CO_2 emissions due to active development activities such as tourism, plantation, and deforestation. This situation might be a driving factor contributing to the increment of CO_2 emissions in the highlands. The higher the CO_2 emissions, the higher the concentration of CO2 in the atmosphere, reducing the moisture within the atmosphere and consequently creating climate change due to the lack of cloud formation and the disruption of the climate phenomenon. This justification is also supported by the humid conditions throughout the year in Malaysia, reflecting the γ values for atmospheric moisture. Almost 90% of CO_2 emissions were from motor vehicles on the roadway of Sabah, especially in urban areas such as Kundasang [41]. The γ values obtained from the correlation of the air temperature for all selected highlands with CO_2 concentration are relatively lower compared to atmospheric moisture, with the γ values ranging from 0.9254 to 1.0153. However, these γ values obtained are significantly related to the CO_2 concentration. According to NASA, continuing a long-term warming trend brought on by human activity, the earth's average surface temperature in 2020 matched with 2016 as the warmest year ever [17]. The highest, lowest, and average temperatures in Cameron Highlands were found to be going up at a

Composite Materials: SEAJCCM2024
Materials Research Proceedings 56 (2025) 30-42

Materials Research Forum LLC
https://doi.org/10.21741/9781644903636-4

rate of 3.8°, 1.8° and 2.8°C in 100 years [3]. Compared to the atmospheric moisture, the air temperature variable is also associated with Malaysia's weather conditions.

Conclusions

In this study, the principal findings show that the concentration of CO_2 emissions is increasing yearly, with ±11% of the percentage of CO_2 increment. The study assimilated the GRA method by considering several variables to estimate the effect of CO2 emission in multiple Malaysian highlands. In conclusion, the performance of the GRA method has successfully determined the factors that have been affected by CO_2 emissions. Atmospheric moisture is the associated factor correlated with the CO2 emissions based on the γ values calculated for all selected highlands, where the atmospheric moisture has been critically affected by the CO_2 emissions compared to the atmospheric temperature. Accordingly, it can be concluded that Cameron Highland has a higher grey relational degree than Genting Highlands, Kundasang, and Kelabit Highland, which is about 2.8125.

Acknowledgements

The authors would like to express a special gratitude to Universiti Teknologi MARA Cawangan Pulau Pinang for providing financial assistance and facilities for this research work. The authors also gratefully acknowledge the use of the Giovanni online data system, developed and maintained by NASA. The Giovanni database provided valuable data that contributed significantly to the analysis presented in this work.

Declaration of Competing Interest

The authors declare that they have no financial interests or personal relationships that could have influenced the work reported in this paper.

References

[1] C.J. Barrow, J. Clifton, N.W. Chan, Y.L. Tan, Sustainable Development in the Cameron Highlands, Malaysia, Malaysia Journal of Environmental Managment 6 (2005) 41–57

[2] M.E. Hamdan, N. Man, S. Md. Yassin, J.L. D'Silva, H.A. Mohamed Shaffril, Farmers Sensitivity Towards Changing Climate in the Cameron Highlands, Agricultural Journal 9 (2014) 120–126

[3] K.W. Tan, P.N. Loh, Climate change assessment on rainfall and temperature in Cameron Highlands, Malaysia, using regional climate downscaling method, Carpathian Journal of Earth and Environmental Sciences 12 (2017) 413–421

[4] W. Kean Fong, H. Matsumoto, C. Siong ho, Y. Fat Lun, Energy Consumption and Carbon Dioxide Emission Considerations in the Urban Planning Process in Malaysia, Planning Malaysia Journal 6 (2008) 99–128. https://doi.org/10.21837/pmjournal.v6.i1.68

[5] K. Alasinrin Babatunde, F. Faizah Said, N. Ghani Md Nor, R. Ara Begum, Reducing Carbon Dioxide Emissions from Malaysian Power Sector: Current Issues and Future Directions (Mengurangkan Pengeluaran Karbon Dioksida dalam Sektor Tenaga di Malaysia: Isu Semasa dan Arah Masa Hadapan), Jurnal Kejuruteraan SI 1 (2018) 59–69. https://doi.org/10.17576/jkukm-2018-si1(6)-08

[6] IEA, Global Energy and CO2 Status Report, Oecd-Iea (2018) 15

[7] T.F. Stoker, D. Qin, G.-K. Plattner, M.M.B. Tignor, S.K. Allen, J. Boschung, A. Navels, Y. Xia, V. Bex, P.M. Midgley, Climate Change 2013 – The Physical Science Basis, Cambridge University Press, 2014

[8] J. Acker, R. Soebiyanto, R. Kiang, S. Kempler, Use of the NASA giovanni data system for geospatial public health research: Example of weather-influenza connection, ISPRS Int J Geoinf 3 (2014) 1372–1386. https://doi.org/10.3390/ijgi3041372

[9] A.I. Prados, G. Leptoukh, C. Lynnes, J. Johnson, H. Rui, A. Chen, R.B. Husar, Access, Visualization, and Interoperability of Air Quality Remote Sensing Data Sets via the Giovanni Online Tool, IEEE J Sel Top Appl Earth Obs Remote Sens 3 (2010) 359–370. https://doi.org/10.1109/JSTARS.2010.2047940

[10] R. Miao, N. Lu, L. Yao, Y. Zhu, J. Wang, J. Sun, Multi-Year Comparison of Carbon Dioxide from Satellite Data with Ground-Based FTS Measurements (2003–2011), Remote Sens (Basel) 5 (2013) 3431–3456. https://doi.org/10.3390/rs5073431

[11] S. Shen, H. Rui, Z. Liu, T. Zhu, L. Lu, S. Berrick, G. Leptoukh, W. Teng, J. Acker, J. Johnson, S.P. Ahmad, A. Savtchenko, I. Gerasimov, S. Kempler, Giovanni: A system for rapid access, visualization and analysis of earth science data online, 86th AMS Annual Meeting (2006) 1–10

[12] R. Miao, N. Lu, L. Yao, Y. Zhu, J. Wang, J. Sun, Multi-Year Comparison of Carbon Dioxide from Satellite Data with Ground-Based FTS Measurements (2003–2011), Remote Sens (Basel) 5 (2013) 3431–3456. https://doi.org/10.3390/rs5073431

[13] I.G. Yashchenko, T.O. Peremitina, Application of the terra MODIS satellite data for environmental monitoring in Western Siberia, International Archives of the Photogrammetry, Remote Sensing and Spatial Information Sciences - ISPRS Archives 41 (2016) 185–187. https://doi.org/10.5194/isprsarchives-XLI-B6-185-2016

[14] T.C. Chang, S.J. Lin, Grey relation analysis of carbon dioxide emissions from industrial production and energy uses in Taiwan, J Environ Manage 56 (1999) 247–257. https://doi.org/10.1006/jema.1999.0288

[15] C. Yuan, D. Wu, H. Liu, Using grey relational analysis to evaluate energy consumption, CO2 emissions and growth patterns in China's provincial transportation sectors, Int J Environ Res Public Health 14 (2017). https://doi.org/10.3390/ijerph14121536

[16] Giovanni: The Bridge Between Data and Science - Eos, (n.d.). https://eos.org/science-updates/giovanni-the-bridge-between-data-and-science (accessed July 20, 2020)

[17] Temperature Change and Carbon Dioxide Change | National Centers for Environmental Information (NCEI) formerly known as National Climatic Data Center (NCDC), (n.d.)

[18] H. Zhang, S. Li, Research on the Factors Influencing CO2 Emission Reduction in High-Energy-Consumption Industries under Carbon Peak, Sustainability (Switzerland) 15 (2023). https://doi.org/10.3390/su151813437

[19] C.J. Barrow, J. Clifton, N.W. Chan, Y.L. Tan, Sustainable Development in the Cameron Highlands, Malaysia, Malaysia Journal of Environmental Managment 6 (2005) 41–57

[20] C.N. Weng, Cameron Highlands Issues and Challenges in Sustainable Development, 2006

[21] M.E. Hamdan, N. Man, S. Md. Yassin, J.L. D'Silva, H.A. Mohamed Shaffril, Farmers Sensitivity Towards Changing Climate in the Cameron Highlands, Agricultural Journal 9 (2014) 120–126

[22] M.H. Mat Zin, Forest encroachment mapping in cameron highlands, using cadastral parcel and remote sensing datasets, Universiti Teknologi Malaysia (2015)

[23] H.D. Tjia, Kundasang (Sabah) at the intersection of regional fault zones of Quaternary age, Bulletin of the Geological Society of Malaysia 53 (2007) 59–66. https://doi.org/10.7186/bgsm53200710

[24] F. Kho, W.E.E. Liang, Levels At Signalized, (2008)

[25] L. Abdullah, H.M. Pauzi, Methods in forecasting carbon dioxide emissions: A decade review, J Teknol 75 (2015) 67–82. https://doi.org/10.11113/jt.v75.2603

[26] A.I. Prados, G. Leptoukh, C. Lynnes, J. Johnson, H. Rui, A. Chen, R.B. Husar, Access, Visualization, and Interoperability of Air Quality Remote Sensing Data Sets via the Giovanni Online Tool, IEEE J Sel Top Appl Earth Obs Remote Sens 3 (2010) 359–370. https://doi.org/10.1109/JSTARS.2010.2047940

[27] U.C.S. Satellite, D. User, The UCS Satellite Database, (n.d.) 1–10

[28] Y. Zhang, N. Kerle, Satellite remote sensing for near-real time data collection, Geospatial Information Technology for Emergency Response (2008) 75–102

[29] AAUSAT-4 – Spacecraft & Satellites, (n.d.). http://spaceflight101.com/spacecraft/aausat-4/ (accessed December 11, 2019)

[30] GOSAT - Earth Online - ESA, (n.d.). https://earth.esa.int/web/guest/missions/3rd-party-missions/current-missions/gosat (accessed December 11, 2019)

[31] RazakSat - eoPortal Directory - Satellite Missions, (n.d.). https://earth.esa.int/web/eoportal/satellite-missions/r/razaksat (accessed December 11, 2019)

[32] C. Yuan, D. Wu, H. Liu, Using grey relational analysis to evaluate energy consumption, $CO2$ emissions and growth patterns in China's provincial transportation sectors, Int J Environ Res Public Health 14 (2017). https://doi.org/10.3390/ijerph14121536

[33] G. Fang, Y. Guo, X. Huang, M. Rutten, Y. Yuan, Combining Grey Relational Analysis and a Bayesian Model Averaging method to derive monthly optimal operating rules for a hydropower reservoir, Water (Switzerland) 10 (2018). https://doi.org/10.3390/w10081099

[34] Y. Fan, L.C. Liu, G. Wu, Y.M. Wei, Analyzing impact factors of $CO2$ emissions using the STIRPAT model, Environ Impact Assess Rev 26 (2006) 377–395. https://doi.org/10.1016/j.eiar.2005.11.007

[35] T.C. Chang, S.J. Lin, Grey relation analysis of carbon dioxide emissions from industrial production and energy uses in Taiwan, J Environ Manage 56 (1999) 247–257. https://doi.org/10.1006/jema.1999.0288

[36] W. Ng Meng, C. Alejandro, A. Abdul Wahab, A Study Of Global Warming In Malaysia, Jurnal Teknologi F 42 (2005) 1–10

[37] S. Wulffraat, C. Greenwood, K.F. Faisal, D. Sucipto, The Environmental Status of Borneo, WWF Report, 2016

[38] Rising carbon emissions | New Straits Times | Malaysia General Business Sports and Lifestyle News, (n.d.)

[39] H.A. Bekhet, N.S. Othman, Impact of urbanization growth on Malaysia $CO2$ emissions: Evidence from the dynamic relationship, J Clean Prod 154 (2017) 374–388. https://doi.org/10.1016/j.jclepro.2017.03.174

[40] D. Ju-Long, Control problems of grey systems, Syst Control Lett 1 (1982) 288–294. https://doi.org/10.1016/S0167-6911(82)80025-X

[41] F.W.L. Kho, P.L. Law, A predictive study: Carbon monoxide emission modeling at a signalized intersection, Journal of Engineering Science and Technology 9 (2014) 1–14

Composite Materials: SEAJCCM2024
Materials Research Proceedings 56 (2025) 43-52

Materials Research Forum LLC
https://doi.org/10.21741/9781644903636-5

Effect of Plant-Based Wastes as a Catalyst for Food Waste Composting on Physio-chemical and Biological Properties

H.S.M. RAIMI[1,a], T.N.H.T. ISMAIL[1,2,b *], R. ALI[3,c], M.Z.M. NAJIB[3,d], Y. YURIZ[1,e] and T. JAIS[4,f]

[1]Department of Civil Engineering Technology, Faculty of Engineering Technology, Universiti Tun Hussein Onn Malaysia, 84600 Muar, Johor, Malaysia

[2]Sustainable Engineering Research Centre, Faculty of Engineering Technology, Universiti Tun Hussein Onn Malaysia, 84600 Muar, Johor, Malaysia

[3]Department of Water & Environment Engineering, Faculty of Civil Engineering, Universiti Teknologi Malaysia, 84600 Muar, Johor, Malaysia

[4]Department of Mechanical & Manufacturing Technology, Kolej Vokasional Muar, 84000 Muar, Johor, Malaysia

[a]diahsmv@gmail.com, [b]hasanah@uthm.edu.my, [c]linda@uthm.edu.my, [d]mohamedzuhaili@utm.my, [e]yasminyuriz@yahoo.com, [f]tahirppu@gmail.com

Keywords: Catalyst, Plant- Based Waste, Food Waste Composting

Abstract. Food waste is not a recent issue; it has long been a serious environmental issue for the ecosystem and human health. This problem becomes more critical as the amount is rising in tandem with the economy and lifestyle. Composting is one of the most sustainable and economical alternative solutions for disposing of food waste directly at source through the development of organic catalysts. Therefore, this research aimed to evaluate the effect of the catalyst developed by local plant-based wastes (lettuce, cucumber, carrot, spinach, iceberg lettuce, pineapple peel, and banana peel) on temperature, moisture content, potential of hydrogen (pH) and microbial population. The result showed that the pH profile for all catalyst resources attained a maturity pH of 3.5, ranging from 1 to 7 days, with the shortest time duration by pineapple (1 day) followed by carrot and cucumber (2 days), green spinach, iceberg lettuce and lettuce (3 days), and banana (7 days). However, the catalyst obtained from cucumber waste has the largest microbial population (lactic acid and yeast). Lettuce, spinach, and pineapple do not produce yeasts or lactic acid bacteria. Furthermore, during the composting process, only the cucumber waste catalyst shows a strong temperature increase into the thermophilic phase on day 1 (46.8°C). In contrast, all others show a temperature rise of less than 40°C. In conclusion, cucumber waste has a huge potential to accelerate food waste composting and enable easy, sustainable, and economical food waste management.

Introduction

Food waste accounts for 45% of municipal solid wastes created, with a total of 27,319 t/day generated in 2020 [1]. Food waste is biodegradable, unused, uneaten, and inedible food that is generated during preparation, distribution and consumption and is disposed of in the food supply chain [1,2]. The increase in food waste in Malaysia from year to year is worrying. Increased purchasing power with economic status and a more advantageous lifestyle have led people to spend more than necessary on food waste. The estimated amount of food waste in 2005 was 4.404 million tons and is increasing to 6.1 million tons per year [3]. Food waste must be managed at a manageable level because it poses a serious threat to the environment and living things, leading to air and water pollution, climate change, soil contamination and a shortage of landfill space [4].

Content from this work may be used under the terms of the Creative Commons Attribution 3.0 license. Any further distribution of this work must maintain attribution to the author(s) and the title of the work, journal citation and DOI. Published under license by Materials Research Forum LLC.

Composite Materials: SEAJCCM2024   Materials Research Forum LLC
Materials Research Proceedings 56 (2025) 43-52   https://doi.org/10.21741/9781644903636-5

In recent years, composting has been considered an acceptable alternative to treating food waste through the decomposition process [5,6]. Food waste decomposition is separated into two techniques: conventional and rapid composting. Conventional composting refers to the decomposition process under natural conditions while microbes or additive material adding is known as rapid composting [7]. The decomposition process under the natural condition approach has lower capital and maintenance costs, easy process design and is effective in minimizing the issue of organic waste [8]. Unfortunately, this method has a major environmental impact because it produces an unpleasant smell during the process [7]. Another problem with the conventional decomposition technique is that it takes longer (around 16-32 weeks) to produce the final product [9,10]. Due to the longer decomposition time, this technique cannot be used to cope with the high amount of food waste in households, as households produce 1.17 kg/day of unavoidable food [1].

According to scholars, using these materials as a catalyst source, i.e. fermentation solution, effective microorganisms, etc., can reduce the decomposition time to only four to eight weeks. There are many studies have been conducted to determine alternative materials as catalysts so that food waste can be decomposed into the final product quickly. The materials that have been used to speed up the food waste decomposition process are yoghurt [11], fermented food (e.g. Tempeh and Tapai) [12], aerated fishpond water [13], kitchen and fruit waste [14,15], commercial effective microorganism [16] and others. Apart from these ingredients, a mixture of fruit and vegetable wastes has also been used to speed up the decomposition process. Among the fruits used include bananas, guava, and sapodilla, while rice, chick-peas, and mustard have been used as vegetable ingredients [33]. However, inappropriate fruit mixtures may cause a decrease in pH to acidic and increase the moisture content. This condition causes decomposing bacteria to not develop effectively due to a shortage of oxygen [40], resulting in the existence of an undesirable stench [41]. In addition, high of moisture content also it will affect the growth of bacteria and fungi in the initial stages of composting [42]. Andraskar [43] reported that a lower initial colony of bacteria and fungi will result in a slower aeration rate of composting process. Therefore, this research attempts to identify the most suitable plant-based waste to develop a catalyst. This catalyst was used to accelerate the food waste decomposition process on a home scale mode.

Materials and Methods

Materials

The materials used in this study are divided into three main components, including catalyst, fermentation bed and simulation food wastes. In this study, a catalyst was used to accelerate the decomposition process of food waste. A fermentation bed is essential to provide a habitat for fermentation microorganisms and enable air ventilation much better, as recommended by Hibino et al. [17]. Last but not least, the simulation food waste is considered as food waste and a problem to be addressed in this research.

Catalyst

The catalyst was prepared by combining plant-based wastes, palm sugar, and distilled water in a ratio of 3:1:10. In this study, seven plant-based wastes were used to increase the microbial activity and accelerate the decomposition process, including lettuce, cucumber, carrot, spinach, iceberg lettuce, pineapple peel, and banana peel. These wastes were collected from a local supermarket in Segamat, Johor. They were then segregated by type and cleaned with water to ensure no residue remained attached to other waste before being cut into 1 cm long and 1 cm wide pieces and placed in different bottles. The palm sugar used in this study was obtained from the sap of the tropical coconut palm or from a palm tree sap (*Arenga pinnata*) [19,20,21] and is usually dark brown in colour, forming a solid after the cooking process. According to Aslanzadeh et al. [18], palm sugar is one of the local carbon sources that can be obtained easily. The contents of sucrose in palm sugar are important as a food resource (energy and nutrient) for microorganism

Composite Materials: SEAJCCM2024   Materials Research Forum LLC
Materials Research Proceedings 56 (2025) 43-52   https://doi.org/10.21741/9781644903636-5

growth in the fermentation process [19]. The palm sugar used in this study was obtained from Pasaraya Mega Segamat, Johor. The palm sugar was then crushed so that it dissolved easily in water. The distilled water is a medium for the growth and metabolism of microorganisms produced during fermentation [22]. In addition, water is used both as a solvent for palm sugar and as an inhibitor. The water was autoclaved at 121 °C for 15 minutes before use.

First, the crushed palm sugar was diluted with 1500 ml of distilled water in a 5L plastic bottle. Then, about 450 g of plant-based waste was placed in this bottle, sealed with a lid, and kept airtight at constant room temperature to allow the fermentative microorganisms to grow. During the process, the plastic bottle is opened regularly to release carbon dioxide until Eco-S has reached the maturation phase. The same process was also used to produce other catalysts, although the types of plant-based waste used were different.

Fermentation Bed

The fermentation bed is prepared with two basic materials, namely one portion of palm biochar and one portion of rice husks (1:1) v/v, in a 21L container with a lid. Palm biochar is ash produced from the combustion of palm waste at 600 ° C. The from this incineration was left in the composting process to settle naturally in the soil and become black soil [23]. Encinar et al. [24] found that palm biochar contains high levels of potassium, which acts as an activator of various metabolic enzymes. In this study, palm biochar was collected from Labis-Segamat, Johor. Rice husk is a by-product derived from the rice milling process. The rice husks used in this study were collected from Southern Jelapang Rice (M) Sdn. Bhd., Muar Johor. Rezagama and Samudro [25] reported that rice husks are used to speed up the composting cycle because they can promote the movement of air and compost. The moisture content in the adequate level of the mixture forms a lump without oozing water when squeezed in the palm hand, as reported by Hibino et al. [17]. The moisture content must be between 40% and 60%, as recommended by Gómez-Brandon et al. [26] and Hubbe et al. [27].

Simulation Food Waste

Simulation food waste was used instead of actual food waste since it was effectively generated by replicating the food waste proximate repercussions of the examination obtained from the Malaysian household unit according to organic acid generation. The simulation food waste used in this study was prepared using a common type of food waste from the kitchen, such as rice, meat, fish, oil, salt, sauces, vegetables, and fruit. The composition of food waste simulated in this study consists of rice, meat, fish, vegetables, fruits, oil, salt and sauces, as shown in Table 1.

Table 1. Simulation of food waste composition.

Composition	Stimulation Food Waste
Rice	1500 g
Meat	125 g
Fish	187.5 g
Vegetables and Fruits	562.5 g
Oil	87.5 g
Salt	12.5 g
Sauces	25 g

Methods

A preliminary test was conducted on random plant-based waste based on the successful catalyst indicator. Next, from the observation of the preliminary test, the plant-based waste which hit the indicator of a successful catalyst was selected for further investigation to identify the main

Composite Materials: SEAJCCM2024
Materials Research Proceedings 56 (2025) 43-52

Materials Research Forum LLC
https://doi.org/10.21741/9781644903636-5

component to produce the catalyst. Several techniques, such as odor test, temperature, pH value, gas emission and microbial population test, were conducted to determine the presence of a catalyst.

Odor and Gas Emission

The odor was observed during the decomposition process by a trained sensory panel according to carefully standardized procedures [28]. A three-member panel included a leader who was selected and trained according to ASTM Committee E-18 on Sensory Evaluation of Materials and Products (1981) [40]. The odor recognition series is used to determine the ability of the panel to identify and describe four (4) odors that indicate the success of Eco-S, as shown in Table 2 [12,29]. Auditory sensory monitoring was used to measure gas emissions. The indicator for evaluating this parameter is "yes" if gas was released and "no" if no gas was produced when the catalyst solution bottle cap was removed. The gas emission is an indication of the success of the catalyst [30].

Table 2. Type of odor emission to indicate the successfully of Eco-S.

Code	Description
Sweet	The aroma of ripe mangos or freshly baked cookies
Sour	The aroma of pickle or sour milk
Rotten	The stench of rotten egg or overripe rubbish
Stinky	The stench of a clogged sewerage pipe or dirty sweaty sock

Microbial Population

The microbial population was assessed by counting the colony-forming units (CFU) on lactic acid and yeast culture media, namely De Man, Rogosa, and Sharpe Agar (MRS) and Dextrose Potato Agar (PDA). A microbial test was conducted on the day of catalyst maturation from day 1 to 7. Culture media in solid form (agar) are used in this study with the preparation of MRS and PDA agar referring to the labels on the culture media container respectively. The agar was poured into a sterile petri dish. 1g of the sample was diluted with sterile distilled water at a dilution ratio of 1/10 before 0.1 ml of the diluted material was added to the middle of an agar medium and spread evenly over the surface using an L-shaped bent glass rod. The agar plate was incubated at 30 °C ± 2 °C mesophilic incubation temperature [28,31]. After 24 hours, the agar plate was removed from the incubator and counted. The colonies on the plate were counted visually using a colony counter. All glassware and pipette tips must be washed, cleaned, and autoclaved at 121°C for 15 minutes before use. All procedures, including agar preparation, sample dilution and plating, were performed in a laminar air floor chamber.

Temperature and pH Value

The temperature and pH of the compost sample were measured daily during the decomposition process using a HI99121 Hanna Direct PH Soil Metre with a temperature probe. The temperature of each compost sample was measured at three different points, and the average readings were recorded as the standard deviation. The pH values of the compost sample were determined at three different points at half the depth of the container [32]. The average of the readings was recorded. Abdullah et al [32] also used a similar method.

Results and Discussion

Odor, Gas Emission, and pH Value

The performance of the plant-based waste solutions is summarized in Table 3. The results show that all plant-based solutions (lettuce, cucumber, carrot, spinach, iceberg lettuce, pineapple peel, and banana peel) made a popping sound when the bottle cap was opened, indicating that all catalyst sources produce gas after one day. This indicates the production of carbon dioxide through the activity of microorganisms [33].

Table 3. Summary of plant observations used to create catalysts from different plant solutions from plant-based waste.

Catalyst Plant-based sources	Odor	Gasses	pH measurement		Capacity to generate microbes
			Day	pH value	
Lettuce	Sour	Yes	3	3.5	Medium
Cucumber	Sweet & sour	Yes	2	3.5	High
Carrot	Sour & stinky	Yes	2	3.5	Medium
Spinach	Sour stinky	Yes	3	3.5	Medium
Iceberg Lettuce	Sweet & sour	Yes	3	3.5	High
Pineapple	Sweet & sour	Yes	1	3.5	High
Banana peel	Sweet & stinky	Yes	7	3.5	Medium

As for the odor criteria, three types of plant-based waste, including cucumber, iceberg lettuce and pineapple peel, produced a sweet and sour odor due to the excellent catalyst quality. Even though the plant-based catalysts from lettuce, carrots, spinach, and banana peels produce a slightly sour odor as opposed to a sweet and sour odor, these four vegetable wastes are considered good catalysts. This is in line with the results of a study by Hibino et al. [17], which states that a catalyst is in a good position if it smells pleasant (e.g. tape or tempeh), but it is contaminated if it smells terrible (rotten).

Next, the pH value of all plant-based catalysts was measured, which had a value of 3.5. However, the time it took for each plant catalyst to reach this value was different. According to the results of the experiment, pineapple reached pH 3.5 the fastest (only one day), followed by cucumber and carrot, which both took two days. Lettuce, spinach, and iceberg lettuce took three days, while the banana solution took the longest (seven days) to reach a pH of 3.5. Fan [12] reported that optimal microbial growth (yeast and lactic acid) occurs when the catalyst has a pH that indicates it can be used in an acidic state (pH <4). In addition, the decrease in pH may also be due to the existence of lactic acid bacteria as the dominant microorganism of the catalyst [34]. In general, in the Takakura method, changes in odour (sweet and sour), gas (emission), and pH (3.5) are indicators of a successfully produced catalyst [29,30,35]. Overall, the odour, gas and pH values indicate that cucumber, pineapple and iceberg lettuce meet the criteria for microbial growth. Consequently, a review of the biological and temperature profile was conducted to confirm the effectiveness of the plant-based catalyst.

Microbial Population

The microbial test was performed on each plant-based catalyst (lettuce, cucumber, carrot, spinach, iceberg lettuce, pineapple, and banana peel) by counting the colony-forming units (CFU) on lactic acid and yeast culture media; this was done before mixing with the composite bed to produce seed compost. The purpose of this test is to determine the initial microbial growth [12]. Table 4 shows the number counts of lactic acid bacteria and yeasts recorded in various plant-based catalysts on the days of maturity. Lactic acid bacteria and yeasts are essential bacteria for most decomposition activities and heat generation during the composting process and prevent pathogens [36,37,38]. It was found that the catalyst prepared from cucumber waste produced the highest lactic acid bacteria (8.96 c.f.u/ml) and yeasts (9.477 c.f.u/ml), followed by the catalyst prepared from the banana peel with 7.9 c.f.u/ml lactic acid bacteria and 8.56 c.f.u/ml yeasts. The catalyst made from carrot waste had a yeast population of 8.57 c.f.u/ml and an insufficient number of lactic acid bacteria (TFTC). In contrast to catalysts made from iceberg lettuce waste, the population of lactic acid bacteria was 6.8 c.f.u/ml and TFTC for yeast. However, no microbial population or TFTC (<30) was found for catalysts made from lettuce, spinach waste and pineapple peel.

Composite Materials: SEAJCCM2024   Materials Research Forum LLC
Materials Research Proceedings 56 (2025) 43-52   https://doi.org/10.21741/9781644903636-5

Overall, only plant-based catalysts made from cucumber waste and banana peel contained microorganisms (lactic acid bacteria and yeast) with the ability to ferment organic matter, inhibit the activity of harmful bacteria, and degrade toxic pollutants and unpleasant odours [12]. According to Sutrisno et al. [14], Xu [36], Xu et al. [37] and Sreenivasan [38], lactic acid bacteria and yeast produce bioactive compounds that accelerate the decomposition and fermentation processes. Nonetheless, the composting process was used to evaluate the performance of microbes produced by each catalyst source.

Table 4. The microbial population of various plant-based catalyst sources at matured days.

Plant-based Catalyst sources	Day of Catalyst matured	Lactic Acid	Yeast	Ability as excellent Catalyst
Lettuce	3	TFTC	TFTC	No
Cucumber	2	8.96	9.477	Yes
Carrot	2	TFTC	8.57	No
Spinach	3	TFTC	TFTC	No
Iceberg Lettuce	3	6.8	TFTC	No
Pineapple	1	TFTC	TFTC	No
Banana	7	7.9	8.56	Yes
Notes: TFTC = too few to count (< 30) Unit: c.f.u/ml				

Temperature

Fig. 1 illustrates the temperature profile of a composting sample inoculated with a plant-based catalyst. In this study, the mesophilic stage occurred at a temperature of 15°C - 44°C. In contrast, the thermophilic temperature range is 45°C- 60°C [29]. The temperature profile of the composting sample inoculated with a catalyst prepared from cucumber waste showed that the temperature increased dramatically to the thermophilic phase on Day 1. After reaching 46.8 °C (thermophilic phase), the temperature gradually decreased and entered the second mesophilic phase (43.9 °C) until the maturation phase, which is close to the ambient temperature (15 °C - 30 °C). Meanwhile, other compost samples, such as lettuce, carrots, spinach, iceberg lettuce, pineapple, banana peels and control(water), did not reach the thermophilic phase. It was hypothesised that a lack of microorganisms at the beginning slowed down the decomposition process, causing the decomposition process to take a long time to complete, or no decomposition occurs [17]. This finding could be related to the microbial count (see Table 3), where the microorganisms (lactic acid bacteria and yeast) in the catalyst prepared from cucumber waste stimulated vigorous microbial activity, resulting in an increase in temperature. The temperatures of carrot, spinach, banana, lettuce, iceberg lettuce, pineapple and control reached peaks at 41.8 °C, 40.8 °C, 40 °C, 39.9 °C, 38.9 °C, 37.8 °C and 33.8 °C respectively, in the mesophilic phase. As a result, only catalysts from cucumber pass from the mesophilic to the thermophilic phase. Bernal et al. [39] reported that the change in composting temperature from the mesophilic to the thermophilic phase indicates microbial activity, which means that the decomposition process is taking place.

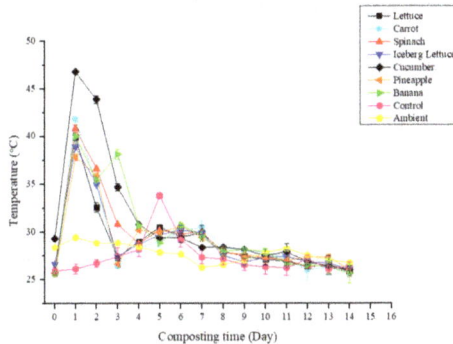

Fig.1. Temperature profile performance of the composting process inoculated by various plant-based solution.

Conclusion

Overall, all plant-based catalysts (lettuce, cucumber, carrot, spinach, iceberg lettuce, pineapple peel, and banana peel) showed acceptable results in terms of odor, gas emission, and pH value even though they reached different times. However, the pineapple peel, carrot and cucumber waste catalysts have a shorter time to reach the maturation stage, which is one and two days. Furthermore, the cucumber catalysts have demonstrated excellent performance in terms of microbial population and temperature profile when inoculated with simulated food waste. It decomposes the simulated food waste faster than other catalyst sources. In addition, the cucumber catalysts reached maturity faster than other plant-based catalysts and control. In conclusion, cucumber was recognized as the best component (microorganism source) for catalyst development due to its physical, chemical, and biological performance for this decomposition study. With this accomplishment, the cucumber catalyst has the potential to be a sustainable, economical, and simple technique for the treatment of food waste at source.

Acknowledgments

This research was supported by ministry of higher education (mohe) through fundamental research grant scheme (FRGS/1/2018/TK10/UTHM/03/3) and Universiti Tun Hussein Onn Malaysia through MDR (vot. Q722).

References

[1] P. Saravanan., S.S. Kumar, C. Ajithan, Eco-friendly practice of utilization of food wastes, International Journal of Pharmaceutical Sciences Innovation 2 (2013) 14-17

[2] W.J. Lim., N.L. Chin., A.Y. Yusof., A. Yahya, T. P. Tee, Food waste handling in Malaysia and comparison with other Asian countries, International Food Research Journal 23 (2016) S1-S6

[3] M. Heikal Ismail., T.I.M. Ghazi., M.H. Hamzah., L.A. Manaf., R.M. Tahir., A. Mohd Nasir, A. Ehsan Omar, Impact of movement control order (MCO) due to coronavirus disease (covid-19) on food waste generation: A case study in klang valley, Malaysia, Sustainability 12 (2020) 8848. https://doi.org/10.3390/su12218848

[4] K. Paritosh., S.K. Kushwaha., M. Yadav., N. Pareek., A. Chawade, V. Vivekanand, Food waste to energy: an overview of sustainable approaches for food waste management and nutrient

recycling, BioMed research international 2017 (2017) 2370927.
https://doi.org/10.1155/2017/2370927

[5] P. Agamuthu, D. Victor, Policy trends of extended producer responsibility in Malaysia, Waste Management & Research 29 (2011) 945-953. https://doi.org/10.1177/0734242X11413332

[6] A.A. Hashim., A.A. Kadir., M.H. Ibrahim., S. Halim, N.A. Sarani, M.I.H. Hassan, N. J. A. Hamid, N.N.H. Hashar, N.F.N. Hissham, Overview on food waste management and composting practice in Malaysia, AIP conference proceedings 2339 (2021) 020181. https://doi.org/10.1063/5.0044206.

[7] H.S. Mohamed Raimi., T.N.H. Tuan Ismail., M.Z. Mohamed Najib, F. Mohamed Yusop, Food waste solution at home: conventional and rapid composting techniques, Supplementary 4 (2020) 1–10. https://doi.org/10.26656/fr.2017.4(S6).016.

[8] W. Namkoong., E.Y. Hwang., J.S. Park., J.Y. Choi, Bioremediation of diesel-contaminated soil with composting, Environmental pollution 119 (2002) 23-31. https://doi.org/10.1016/S0269-7491(01)00328-1.

[9] Information on
https://issuu.com/anwar_townplan/docs/laporan_panduan_pelaksanaan_pengkom

[10] S.I. Sharifah Norkhadijah, Application of effective microorganism (EM) in food waste composting: A review, Asia Pacific Environmental and Occupational Health Journal 2 (2016) 37-47

[11] F.A. Nuzir., S. Hayashi, K. Takakura, Takakura composting method (TCM) as an appropriate environmental technology for urban waste management, International Journal of Building, Urban, Interior and Landscape Technology (BUILT) 13 (2019) 67-82. https://doi.org/10.14456/built.2019.6

[12] Y. Van Fan., C.T. Lee., C.W. Leow., L.S. Chua, M.R. Sarmidi, Physico-chemical and biological changes during co-composting of model kitchen waste, rice bran and dried leaves with different microbial inoculants, Malaysian Journal of Analytical Sciences 20 (2016) 1447-1457. http://doi.org/10.17576/mjas-2016-2006-25

[13] S.N.B.B. Khalib., Z.I. Azura, T.T. Nuraiti, Mini Review: Environmental Benefits of Composting Organic Solid Waste by Organic Additives, Bulletin of Environmental Science and Sustainable Management 2 (2014) 1-7. https://doi.org/10.54987/bessm.v2i1.40

[14] E. Sutrisno., B. Zaman., I.W. Wardhana., L. Simbolon, R. Emeline, Is Bio-activator from Vegetables Waste are Applicable in Composting System?, IOP Conference Series: Earth and Environmental Science 448 (2020) 012033. https://doi.org/10.1088/1755-1315/448/1/012033

[15] N. Ismail., M. Firdaus., M.K. Ismail, M.N. Samat, Effects of Vermicomposts and Home-Made Effective Microorganism Based Fertilizers on The Growth of Luffa acutangula (L.) Roxb, 3rd International Biotechnology and Biodiversity Conference & Exhibition (BIOJOHOR 2012), 2012

[16] A. Javaid, R. Bajwa, Field evaluation of effective microorganisms (EM) application for growth, nodulation, and nutrition of mung bean, Turkish Journal of Agriculture and Forestry 35 (2011) 443-452. https://doi.org/10.3906/tar-1001-599

[17] K. Hibino, Operation manual for small-to-medium scale compost centres using the takakura composting method, Institute for Global Environmental Strategies (2020) 3

[18] S. Aslanzadeh., K. Kho, I. Sitepu, An evaluation of the effect of Takakura and effective microorganisms (EM) as bio activators on the final compost quality, IOP Conference Series:

Composite Materials: SEAJCCM2024
Materials Research Proceedings 56 (2025) 43-52

Materials Research Forum LLC
https://doi.org/10.21741/9781644903636-5

Materials Science and Engineering 742 (2020) 012017. https://doi.org/10.1088/1757-899X/742/1/012017

[19] R. Rusdianasari., A. Syakdani., M. Zaman., F.F. Sari., N.P. Nasyta, R. Amalia, Utilization of eco-enzymes from fruit skin waste as hand sanitizer, AJARCDE (Asian Journal of Applied Research for Community Development and Empowerment) 5 (2021) 23-27. https://doi.org/10.29165/ajarcde.v5i3.72

[20] A. Apriyantono., A. Aristyani., Y. Lidya., S. Budiyanto, S. T. Soekarto, Rate of browning reaction during preparation of coconut and palm sugar, International Congress Series 1245 (2002) 275-278. https://doi.10.1016/S0531-5131(02)00882-8

[21] P. Gervais, P. Molin, The role of water in solid-state fermentation, Biochemical Engineering Journal 13 (2003) 85-101. https://doi.10.1016/S1369-703X(02)00122-5

[22] I. Zahrina, Utilization of Palm Ash and Palm Shell as a Source of Silica in the Synthesis of ZSM-5 from Natural Zeolite, Journal of Science and Technology 6 (2007) 31-34

[23] J.M. Encinar., J.F. González., A. Rodríguez-Reinares, Biodiesel from used frying oil. Variables affecting the yields and characteristics of the biodiesel, Industrial & Engineering Chemistry Research 44 (2005) 5491-5499. https://doi.org/10.1021/ie040214f

[24] A. Rezagama, G. Samudro, Optimization study of takakura with the addition of husk and bran, Journal of precipitation 12 (2015) 66-70. https://doi.org. 10.14710/presipitasi.v12i2.66-70

[25] M. Gómez-Brandón, C. Lazcano, J. Domínguez, The evaluation of stability and maturity during the composting of cattle manure, Chemosphere 70 (2008) 436-444. https://doi.org.10.1016/j.chemosphere.2007.06.065

[26] M.A. Hubbe., M. Nazhad, C. Sanchez, Composting as a way to convert cellulosic biomass and organic waste into high-value soil amendments: a review, BioResources 5 (2010) 2808-2854. https://doi.org/10.15376/biores.5.4.2808-2854

[27] K.A. Ismail., H.M. S. El-Din., S.M. Mohamed., A.B.M.A. Latif, M. A. M. Ali, Monitoring of physical, chemical, microbial and enzymatic parameters during composting of municipal solid wastes: a comparative study, Journal of Pure and Applied Microbiology 8 (2013) 211-224

[28] A.M. Musa., C.F. Ishak., D.S. Karam, N. Md Jaafar, Effects of fruit and vegetable wastes and biodegradable municipal wastes co-mixed composts on nitrogen dynamics in an Oxisol, Agronomy 10 (2020) 1609. https://doi.org/10.3390/agronomy10101609

[29] G.H. Ying, M.H. Ibrahim, Local Knowledge in Waste Management: a study of Takakura home method, JECET 2 (2013) 528-533

[30] A.E. Yousef, C. Carlstrom, Food microbiology: A laboratory manual, John Wiley & Sons, 2003. https://doi.org/10.1186/2251-7715-2-3

[31] N. Abdullah., N.L. Chin, M.N. Mokhtar, F.S. Taip, Effects of bulking agents, load size or starter cultures in kitchen-waste composting, International Journal of Recycling of Organic Waste in Agriculture 2 (2013) 1-10. https://doi.org/10.1186/2251-7715-2-3

[32] S.N.B. Zailani, Moisture content influence on composting parameters and degradation kinetic models in an aerated closed system, Doctoral dissertation, PhD thesis, Universiti Teknologi Malaysia, 2018.

[33] D.W. Widjajanto, E.D. Purbajanti, U.C. Sumarsono, The role of local microorganisms generated from rotten fruits and vegetables in producing liquid organic fertilizer, J Applied Chem. Sci. 4 (2017) 325-329. http://doi.org/10.22341/jacson.00401p325

[34] Y. Van Fan., C.T. Lee, J.J. Klemeš., L.S. Chua., M.R. Sarmidi, C.W. Leow, Evaluation of Effective Microorganisms on home scale organic waste composting, Journal of Environmental Management 216 (2018) 41-48. https://doi.org/10.1016/j.jenvman.2017.04.019

[35] H.L. Xu, Effects of a microbial inoculant and organic fertilizers on the growth, photosynthesis and yield of sweet corn, Journal of crop production 3 (2001) 183-214. https://doi.org/10.1300/J144v03n01_16

[36] H.L. Xu, M.A.U. Mridha, Effects of organic fertilizers and a catalyst on leaf photosynthesis and fruit yield and quality of tomato plants, Journal of Crop production 3 (2001) 173-182

[37] E. Sreenivasan, Evaluation of effective microorganisms technology in industrial wood waste management, International Journal Advanced Engineering Technology 21 (2013) 22

[38] M.P. Bernal., J.A. Alburquerque, R. Moral, Composting of animal manures and chemical criteria for compost maturity assessment. A review, Bioresource technology 100 (2009) 5444-5453. https://doi.org/10.1016/j.biortech.2008.11.027

[39] ASTM Committee E-18 on Sensory Evaluation of Materials and Products. Guidelines for the selection and training of sensory panel members 1981; 758. ASTM International.

[40] P.V. Pipiana, S. Sunarsih, Y. Pratiwi, Comparison of the effectiveness of MOL bioactivator banana peel (Musa paradisiaca L.) and EM4 in aerobic composting of Strobilanthes cusia leaf waste, Serambi Engineering Journal 9 (2024) 7978-87. https://jse.serambimekkah.id/index.php/jse/article/view/90

[41] C. Sundberg, E. Azzi, Biochar sustainability, in: J. Lehmann, S. Joseph (3rd ed.), *Biochar for Environmental Management,* Abingdon: Routledge, 2024, pp. 785-804

[42] A. Sharma, R. Soni, S.K. Soni, Decentralized in-vessel composting: an efficient technology for biodegradable solid waste management, *Biomass Convers Biorefin.* 14 (2023) 23775-23792. https://doi.org/10.1007/s13399-023-04508-y

[43] J. Andraskar, S. Yadav, A. Kapley, Challenges and control strategies of odor emission from composting operation, *Appl Biochem Biotechnol.* 193 (2021) 2331-2356. https://doi.org/10.1007/s12010-021-03490-3

Composite Materials: SEAJCCM2024
Materials Research Proceedings 56 (2025) 53-68

Materials Research Forum LLC
https://doi.org/10.21741/9781644903636-6

A Multicriteria Assessment of the Environmental Effects of Mass Solar Photovoltaic Farms in Pulau Burung, Pulau Pinang

Siti Isma Hani Ismail[1,a] *, Shanker Kumar Sinnakaudan[1,b], Noorsuhada Md Nor[1,c], Hun Beng Chan[1,d], Zulfairul Zakaria[1,e], Farah Ezlyn Muhamad Norazlan[1,f]

[1]Civil Engineering Studies, College of Engineering, Universiti Teknologi MARA, Cawangan Pulau Pinang, Permatang Pauh Campus, Malaysia

[a]sitiismai@uitm.edu.my, [b]drsshanker@gmail.com, [c]ida_nsn@uitm.edu.my, [d]hunbeng@uitm.edu.my, [e]zulfairul@uitm.edu.my, [f]farahezlyn@gmail.com

Keywords: Large-scale Solar, Renewable Energy, Environmental Impact Screening

Abstract. The proliferation of mass solar photovoltaic (LSSPV) farming has been catalyzed by the escalating costs of fossil fuels and the imposition of carbon pricing, rendering solar energy an increasingly compelling alternative. Nevertheless, significant impediments, including land scarcity, ecological ramifications, and stringent regulatory frameworks, continue to constrain its widespread implementation. This study undertakes a rigorous evaluation of the Environmental Impact Assessment (EIA) for an LSSPV project in Pulau Pinang, Malaysia, utilizing the Analytic Hierarchy Process (AHP) within a Multi-Criteria Decision-Making (MCDM) paradigm to systematically assess its environmental implications. The investigation elucidates a spectrum of critical environmental concerns, encompassing land-use transformations, biodiversity attrition, hydrological perturbations, and the latent risks of chemical contamination. 10 stakeholder analyses delineate ecological degradation and terrestrial alterations as paramount issues, succeeded by hydrological destabilization attributable to surface runoff and the potential hazards posed by chemical constituents within photovoltaic components. The findings substantiate the implementation of mitigation strategies, including rigorous environmental surveillance and the adoption of sustainable land management practices, to attenuate deleterious consequences. Empirical survey data reveal a dichotomy of perspectives, wherein 29% of stakeholders contend that residential proximity to LSSPV installations engenders substantive risks to both environmental integrity and public health, whereas a majority of 71% refute such assertions, perceiving no discernible environmental or health-related perils. The AHP model serves as an intricate evaluative mechanism, synthesizing heterogeneous stakeholder insights to enhance the optimization of sustainability

Introduction

The global shift toward renewable energy sources is becoming increasingly urgent as the world grapples with the growing challenges of climate change and environmental degradation. Among these renewable sources, solar energy stands out as a promising solution, offering clean and sustainable electricity generation. One of the most efficient methods for capturing solar energy is using photovoltaic (PV) technology [1]. In Malaysia, PV is a dynamic and rapidly advancing sector, propelled by the nation's dedication to renewable energy and sustainability. The Malaysian government has introduced a range of initiatives designed to encourage the widespread adoption of solar PV systems, with the goal of minimizing carbon emissions and decreasing dependence on fossil fuels. Malaysia has made Initiatives such as the Malaysia Building Integrated Photovoltaic (MBIPV) and the Feed-in Tariff [2]. These efforts have led to a growing market for PV systems in the country, with the potential for solar PV to become the most competitive renewable energy source by 2050 [3]. The potential benefits of these solar farms in mitigating greenhouse gas emissions and reducing reliance on fossil fuels are well-documented, their environmental impacts

Content from this work may be used under the terms of the Creative Commons Attribution 3.0 license. Any further distribution of this work must maintain attribution to the author(s) and the title of the work, journal citation and DOI. Published under license by Materials Research Forum LLC.

remain an area of concern. The Environmental Impact Assessment (EIA) is an essential tool for assessing the potential environmental effects of proposed projects prior to their implementation which provide crucial data that enable policymakers and planners to make well-informed decisions regarding the installation and operation of PV plants, thereby ensuring sustainable development [4]. The lifecycle assessment (LCA) of large-scale solar PV plants in Malaysia indicates an emission rate of 0.0309 kgCO2eq/kWh, with the construction phase being the primary contributor to emissions, owing to infrastructure development and the manufacturing of PV modules [5]. Dye-sensitized solar cells (DSSC) exhibit a greenhouse gas (GHG) emission rate of 70.52 gCO2-eq/kWh, primarily driven by electricity usage during the manufacturing process [6]. EIA also inform the development of strategies to mitigate negative impacts, such as optimizing design, utilizing innovative materials, and improving recycling processes. Evidently, post-consumer management of PV components, particularly through recycling, can substantially mitigate environmental impacts. Recycling has the potential to reduce greenhouse gas emissions by as much as 42% [7].

Large Scale Solar Scheme

LSS initiatives in Malaysia play a pivotal role in the nation's strategy to enhance renewable energy adoption and decrease dependence on fossil fuels. These programs are essential for fulfilling Malaysia's commitments under the Paris Agreement and for achieving its national renewable energy goals. The government has been proactively advocating for solar energy, with Large Scale Solar (LSS) being the most prominent program overseen by the Sustainable Energy Development Authority (SEDA) and the Energy Commission (EC) [8]. EC serves as the implementing agency for the LSS scheme, a competitive bidding initiative aimed at lowering the Levelized Cost of Energy (LCOE) in the development of large-scale solar (LSS) photovoltaic plants. The scheme in which a bidder is initially allowed to bid for a maximum capacity of 50MWac and may submit no more than three bids, namely LSS1, LSS2, and LSS3. LSS projects play a crucial role in reducing greenhouse gas emissions, supporting Malaysia's objective of achieving a 45% reduction by 2030 [9]. The integration of LSS with Battery Energy Storage Systems (BESS) is also vital for optimizing solar power generation and ensuring grid stability [10]. Research has demonstrated that various types of solar PV panels and battery technologies can be optimized for cost-effective energy storage and efficient utilization [11]. Table 1 shows the first LSS1 project for commercial operation in 2017. The performance of LSS plants is typically assessed using the Performance Ratio (PR), a critical indicator of plant efficiency. Research has shown that the actual PR can surpass target benchmarks, suggesting greater-than-expected energy production. However, the use of incorrect PR formulas may result in energy shortfalls, highlighting the importance of precise performance evaluations [12]. Additionally, incorporating reactive power compensation techniques can help stabilize voltage fluctuations, improve grid integration, and minimize system losses [13].

Composite Materials: SEAJCCM2024
Materials Research Proceedings 56 (2025) 53-68

Materials Research Forum LLC
https://doi.org/10.21741/9781644903636-6

Table 1. The first LSS1 project for commercial operation in 2017

Shortlisted Bidder for Package P3 (30MW to 50MW) in Peninsular Malaysia	Export Capacity (MW)	Location
Tenaga Nasional Bhd	50	Sepang, Selangor
Mudajaya Corp Bhd	49	Sungai Siput, Kuala Kangsar, Perak
Consortium Malakoff Corp Bhd and DRB-Hicom Environmental Services Sdn Bhd	50	Tanjumg Malim, Perak
Consortium Synergy Generated Sdn Bhd, Scomi Group Bhd and Lembaga Tabung Angkatan Tentera	30	Bandar Sungai Petani, Kuala Muda, Kedah

The Environmental Impact Assessment

EIA is a systematic and comprehensive process designed to assess the potential environmental, social, and economic consequences of a proposed project before its execution. This evaluation facilitates the identification of adverse effects and the formulation of mitigation strategies while optimizing beneficial outcomes, thereby promoting sustainable development and informed decision-making [14]. It is essential to recognize that the EIA process does not inherently ensure that a project will be altered or denied, even when significant environmental concerns are identified. The effectiveness of EIA is analysed through procedural, substantive, transactive, and normative dimensions. While procedural effectiveness is widely researched, there is a growing need for further investigation into the transactive dimension, particularly in terms of cost efficiency. Emerging multidimensional studies underscore the importance of understanding the interconnections between these dimensions [15]. Fig. 1 delineates the various stages of the EIA process, along with their respective roles in assessing and addressing potential environmental impacts while Table 2 outlines the primary stages of the EIA process along with their respective details. Large-scale solar (LSS) installations have both beneficial and detrimental environmental impacts. While they play a crucial role in reducing carbon emissions and advancing renewable energy, they also present specific environmental challenges that require attention and mitigation. LSS installations demand vast areas of land, which can result in habitat disruption and changes in land use. This issue becomes particularly critical when forests are cleared for solar farms, leading to an increase in greenhouse emissions [16]. Large-scale PV deployments have the potential to alter local temperature and airflow patterns; however, studies indicate that the negative effects on air temperature and urban heat islands are minimal [17]. Additionally, the decommissioning phase can lead to waste generation and environmental degradation if not managed properly [18].

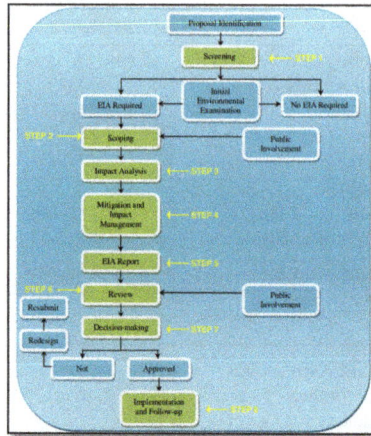

Fig. 1. Stages of the EIA process [19].

Table 2. The primary stages of the EIA process along with their respective details.

Stage	Details
Screening	Determining whether an EIA is necessary or not
Scoping	Sets the boundaries for the EIA and involves stakeholder consultation to ensure all relevant concerns are considered
Investigation and preparing of report	Detailed analysis of the potential environmental impacts identified during scoping. It includes predicting and evaluating the significance of impacts and exploring alternatives to mitigate negative effects
Environmental Management Plan (EMP)	EMP is developed to outline how the identified impacts will be managed and mitigated. This plan includes monitoring and management strategies to ensure compliance with environmental standards
Review	The EIA Report and development application should be made publicly (including through electronic methods), relevant stakeholders and the public must be given the chance to comment on it
Decision	Before deciding whether to approve consent for the development, the competent authority must consider the EIA Report as well as any comments made on it. It is necessary to make the decision statement publicly Environmental Conservation Department

Multi-Criteria Decision-Making

MCDM methods are vital tools for decision-makers who must assess and select from multiple alternatives based on a range of criteria. It is an essential tool in EIA for evaluating and ranking various environmental components and project alternatives with the Analytic Hierarchy Process (AHP) being the most applied method. AHP operates on the principle of breaking down a problem into a hierarchical structure of more manageable sub-problems, each of which can be analysed separately [20]. The process involves conducting pairwise comparisons of elements, which are then used to generate ratio scale estimates that reflect the relative priorities of criteria and alternative solutions [21]. This method incorporates both subjective and objective factors in decision-making, making it a flexible and effective tool for tackling complex decisions [22]. The fuzzy analytic hierarchy process (FAHP) is an adaptation of AHP that integrates fuzzy logic to address uncertainty and imprecision in decision-making. Although FAHP may yield different solutions compared to AHP, it does not necessarily provide a superior quality of solutions [23]. Fig. 2 shows the analytical hierarchy process flowchart.

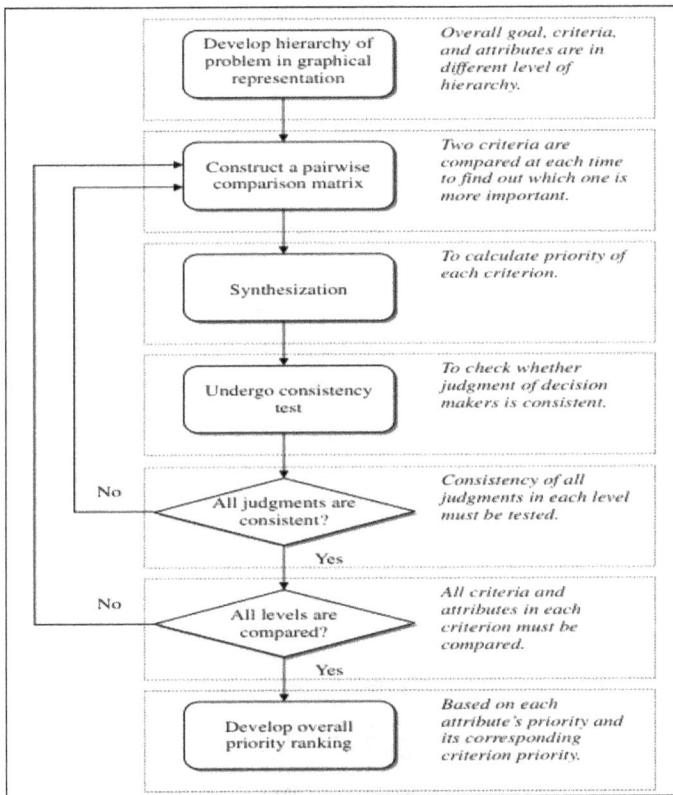

Fig. 2. The analytical hierarchy process flowchart [24].

Methodology

This study employed a quantitative research methodology, utilizing structured, in-depth interviews to collect data from primary sources. Prior to conducting the interviews', meticulously designed questionnaires were developed to target key industry professionals, including academicians, architects, quantity surveyors, engineers, safety officers, solar facility managers, and regulatory authorities. The primary objective of this research was to evaluate the challenges and

Composite Materials: SEAJCCM2024   Materials Research Forum LLC
Materials Research Proceedings 56 (2025) 53-68   https://doi.org/10.21741/9781644903636-6

environmental implications associated with large-scale solar photovoltaic (LSSPV) projects and to formulate strategic measures to mitigate potential risks. Data collection and analysis were executed using the Multi-Criteria Decision-Making (MCDM) framework and the Analytic Hierarchy Process (AHP), ensuring a systematic and rigorous approach to information organization and facilitating informed, evidence-based decision-making. The research approach integrates both desk study and field study within its framework. The procedure and sequence of tasks involved in this study are illustrated in Fig 3. As part of the desk study, Environmental Impact Assessment (EIA) evaluations must be reviewed to determine whether an EIA is required for the construction plan. The data collection process during the field study involves conducting a site survey, interpreting the area using Google Street View and Google Earth, and analyzing the site. Data was gathered through a site investigation and interviews with stakeholders and relevant authorities from the local government. To analyze the data collected, the Analytic Hierarchy Process (AHP) can be employed.

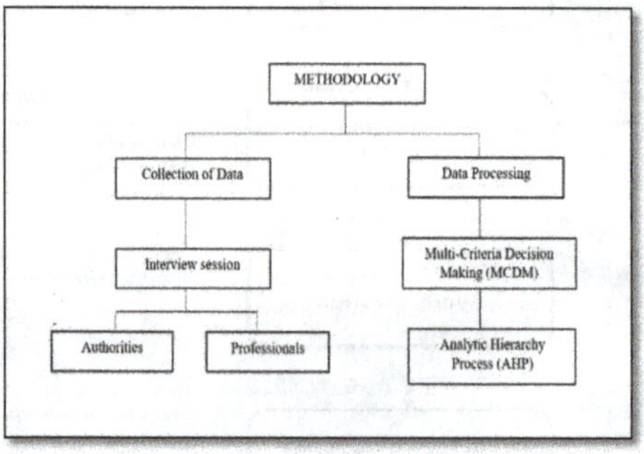

Fig. 3. Flowchart of Data Collection.

Location of Research Area

The designated study area encompasses approximately 53 acres in Pulau Burung, Pulau Pinang. Fig. 4 illustrates the site location and its surrounding environment, emphasizing the potential effects of the solar energy project on nearby residential communities, solid waste facilities, educational institutions, and other critical infrastructure while Fig. 5 indicates its legend.

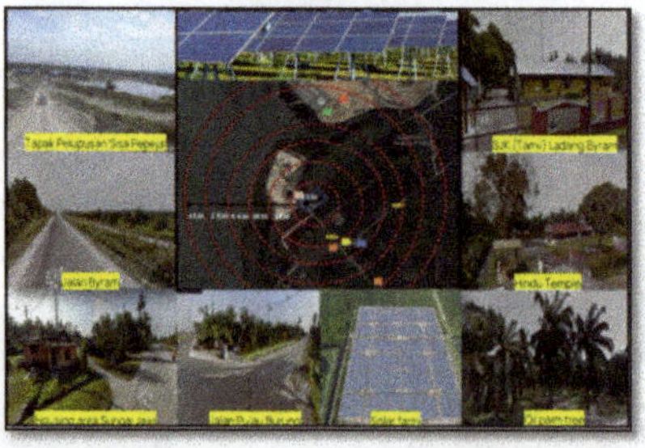

Fig. 4. Location of research area within 5km radius including nearest facilities involved.

Courier station	Temple
Factory	School
Solid waste disposal	Residential Area

Fig. 5. Legend.

Composite Materials: SEAJCCM2024
Materials Research Proceedings 56 (2025) 53-68

Materials Research Forum LLC
https://doi.org/10.21741/9781644903636-6

Data Collection and Analysis

The data for this study was gathered through comprehensive interviews with 10 stakeholders who possess expertise in the field. The participants included academicians, consultants, and representatives from key local authorities, including the Local Council, Malaysian Public Works Department (JKR), Department of Environment (JAS), and Department of Irrigation and Drainage (JPS). These professionals were selected due to their direct involvement in the project's development, ensuring the relevance and reliability of the collected data. As indicated in Table 3, each expert was assigned a unique code (P1, P2, P3, etc.). Throughout the interviews, they provided insights into the potential environmental impacts of large-scale solar photovoltaic (LSSPV) projects. The collected responses were subsequently analyzed using the Analytic Hierarchy Process (AHP) in Excel to systematically evaluate and interpret the findings.

Table 3. List of respondents who specialist in the filed study to provide professional output

No.	Code	Department
1	P1	Engineering Advisor
2	P2	Engineering Advisor
3	P3	Expert Panel
4	P4	Engineering Officer (MBSP)
5	P5	Engineering Officer (MBSP)
6	P6	Building Officer (MBSP)
7	P7	Building Officer (MBSP)
8	P8	Engineer (JKR)
9	P9	Officer (JAS)
10	P10	Officer (JPS)

Interview Session

The interview sessions with local authorities, advisors, and experts were conducted to obtain technical insights and professional opinions regarding the environmental impacts of large-scale solar photovoltaic (LSSPV) energy projects. The discussions centred on understanding the underlying causes of these environmental impacts and exploring effective strategies to mitigate adverse effects on the environment. To collect their insights, the experts were given a questionnaire or participated in an interview session. Their responses provided valuable information on the environmental implications of large-scale solar energy projects and their professional knowledge on the topic. Fig 6 is a sample questionnaire used during interviews, showcasing how participants shared their opinions for this study.

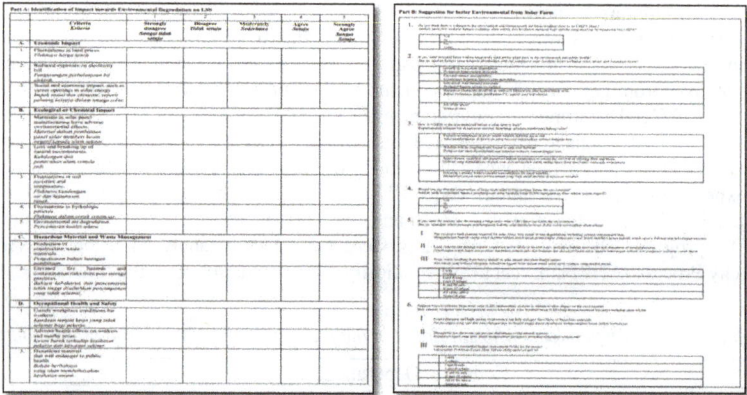

Fig. 6. Sample Questionnaire.

Analytical Hierarchy Process (AHP) using Excel Template

To facilitate data analysis using the AHP template, MS Excel 2013 is required. The workbook consists of 20 input sheets dedicated to pairwise comparisons, supplemented by additional sheets for consolidating judgments and summarizing results. Moreover, it includes reference tables encompassing the randomness index, judgment ranges, and geometric consistency index (GCI) limit ranges. Additionally, the template incorporates a specialized sheet for solving the eigenvalue problem through the eigenvector method (EVM), ensuring a systematic and structured approach to decision analysis. The AHP template accommodates a maximum of 10 criteria and allows participation from up to 20 decision-makers. Fig. 7 presents the results obtained using the eigenvector method (EVM), which calculates the weights and associated errors for each criterion. Additionally, the figure includes a verification field to assess the accuracy of EVM computations. The necessary number of iterations is indicated in the column labelled "Iterations." To ensure the reliability of the results, the "EVM check" value should be as close to zero as possible.

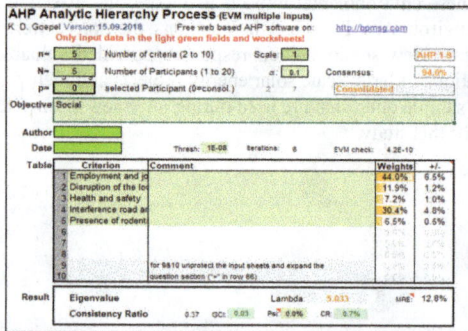

Fig. 7. Sample results obtained using the eigenvector method (EVM).

Result and Analysis

The study's findings are summarized in Table 6 and Fig. 8. Table 4 outlines the impact criteria, which were systematically analyzed using the Analytic Hierarchy Process (AHP) methodology. The table categorizes three principal components—Economic Impact, Ecological and Chemical

Composite Materials: SEAJCCM2024   Materials Research Forum LLC
Materials Research Proceedings 56 (2025) 53-68   https://doi.org/10.21741/9781644903636-6

Impact, and Occupational Health and Safety (OHS) Impact—alongside their respective weightages and rankings for each criterion. This structured analysis provides a comprehensive evaluation of the environmental considerations associated with large-scale solar photovoltaic (LSSPV) projects.

Table 4. The impact criteria weight after of the study.

Criteria		Sub-criterion	Weight	Impact	Ranking
EI		**Economic Impact**			
	EI1	Fluctuations in land prices.	0.264	-	3
	EI2	Reduced expenses on electricity bill.	0.436	+	1
	EI3	Social and economic growth	0.300	+	2
EC		**Ecological and Chemical Impact**			
	EC1	Materials hazard in solar panel manufacturing	0.120	-	5
	EC2	Loss and breaking up of natural environments	0.234	-	4
	EC3	Fluctuations in soil moisture and temperature	0.241	-	2
	EC4	Fluctuations in hydrologic patterns	0.264	-	1
	EC5	Environmental air degradation	0.090	-	6
	EC6	Reduce carbon emissions	0.244	+	3
OSH		**Occupational Health and Safety Impacts**			
	OSH1	Unsafe workplace conditions	0.198	-	4
	OSH2	Adverse health effects on workers and nearby areas	0.166	-	5
	OSH3	Hazardous material endangers to public health	0.229	-	**2**
	OSH4	Production of construction-related solid waste	0.307	-	1
	OSH5	Heightened danger of fires and environmental pollution resulting from improper material storage	0.212	-	3

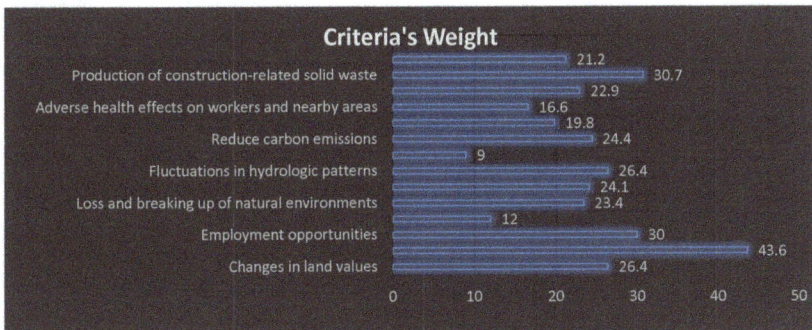

Fig. 8. Chart for each criteria's weight of this study.

Composite Materials: SEAJCCM2024   Materials Research Forum LLC
Materials Research Proceedings 56 (2025) 53-68   https://doi.org/10.21741/9781644903636-6

According to the survey findings, 29% of stakeholders concur that residing near a large-scale solar (LSS) photovoltaic site may pose risks to both environmental and human health. Conversely, the majority, accounting for 71%, maintain that living near an LSS does not present any significant threats to either the environment or public well-being. Solar inverters and photovoltaic (PV) technology are generally considered environmentally friendly compared to fossil fuels [24]. The survey further investigated potential environmental and health hazards associated with residing near a large-scale solar (LSS) photovoltaic facility. Key concerns included habitat loss, an elevated risk of cancer, the presence of toxic chemicals used in the photovoltaic manufacturing process— such as hydrochloric acid—and exposure to hazardous materials. A summary of these findings is presented in Table 5. Solar panels have the potential to modify microclimatic conditions, including temperature and humidity levels, which in turn can impact local biotic communities. For example, the area beneath solar panels typically experiences lower temperatures and higher relative humidity compared to the surrounding environment. These variations can influence species composition and abundance, potentially altering local ecosystems [25].

As presented in Table 6, three fundamental aspects were identified through a comprehensive assessment involving surveys, field observations, and stakeholder interviews. These aspects provide critical insights into the environmental impact associated with the construction of large-scale solar photovoltaic (LSSPV) projects in Malaysia. The impact includes flora and fauna, the broader environmental consequences of LSSPV development, and proposed strategies for enhancing environmental sustainability in solar energy projects. Each category was analyzed using data from surveys. The results were converted into percentages to highlight stakeholder concerns and priorities. Implementing sustainable site-selection criteria that prioritize habitat protection and consider the cumulative impacts of solar facilities is crucial. This includes avoiding areas with high ecological value and focusing on urban or degraded lands to protect biodiversity, they should implement measures such as habitat relocation, landscape modifications and conservation planning [26]. Additionally, EIA systematically assesses the potential environmental consequences of proposed projects, ensuring that development aligns with sustainable development goals (SDG). By incorporating environmental considerations into planning and decision-making processes, EIA facilitates the early identification and implementation of mitigation measures, thereby minimizing adverse environmental effects [27].

Table 5. The risk on environmental and health damage.

Criteria	Percentage (%)
Escalation of natural habitat loss	36
Rise in cancer susceptibility	24
Interaction with harmful materials	20
Hazardous chemicals involved in solar cell fabrication, like hydrochloric acid.	20

Composite Materials: SEAJCCM2024
Materials Research Proceedings 56 (2025) 53-68

Materials Research Forum LLC
https://doi.org/10.21741/9781644903636-6

Table 6. The impact percentage during LSS construction.

Criteria	Percentage (%)
Impact on flora and fauna	
Restrict development in areas where wildlife habitats are at risk	59
Habitat displacement for wildlife	11
Repositioned, modified, and preserved habitat landscapes to ensure the survival of existing flora and fauna	17
Initiating a project without proper consideration for local wildlife	13
Impact on environment	
The extensive land clearing required for solar farms may result in soil degradation, including erosion and nutrient loss	52
Land clearing can damage natural vegetation and wildlife in several ways, including habitat destruction and disruption of rainfall patterns.	39
Toxic waste resulting from heavy metals in solar panels and their fragile nature.	9
Suggested Enhancements for LSS to Benefit the Environment	
Proper planning and high-quality maintenance can help mitigate the effects of hazardous materials.	40
Thoughtful site decisions can prevent disruptions to the natural scenery.	25
Conduct an Environmental Impact Assessment (EIA) for the project.	35

Mitigation Measures

This study emphasizes the necessity of implementing mitigation measures to minimize the environmental impact of large-scale solar photovoltaic (LSSPV) systems. Table 7 below presents the proposed mitigation strategies identified in this research.

Table 7. Mitigation measure on environmental impact.

Sub-criterion	Mitigation Measures
ECONOMIC IMPACT	
Changes in land values	Positive impact.
Cost saving payment on electric bill	Positive impact.
Employment opportunities	Positive impact.
ECOLOGICAL AND CHEMICAL IMPACT	
Materials hazard in solar panel manufacturing	A comprehensive mitigation system should be established to minimize any potential adverse environmental impacts arising from development activities while ensuring the safeguard of ecological integrity.
Loss and breaking up of natural environments	Construction phase • Avoid the forest's critical habitat region. • Plan daily activities to minimize disturbances to wildlife during critical periods, such as mating seasons. • Implement proper waste disposal methods to prevent construction debris from harming aquatic ecosystems. • Install visual warning markers at regular intervals on transmission lines to reduce the risk of bird collisions.

- Remove trees in stream buffer zones that may encroach upon conductor clearance areas within three to four years

Operation

- Minimize human and vehicular activities to reduce environmental disturbances.
- Restore and conserve as much of the natural landscape as possible.
- Use only the necessary number of lights to minimize disruption to migratory birds and endangered species while ensuring safety and security.
- Report active raptor nests found on transmission line structures to wildlife organizations. Remove abandoned nests that no longer contain eggs, with proper coordination with wildlife agencies for relocation or disposal

Decommissioning

- Maximize the use of reclaimed land to minimize habitat loss and fragmentation.
- Restore degraded areas by replenishing topsoil collected at the project's inception or during decommissioning.

Fluctuations in soil moisture and temperature

- When constructing roads, consider local climate conditions, soil moisture levels, and erosion risks to prevent soil degradation and alterations in surface water runoff.
- Implement protective soil cover and surrounding vegetation to support the growth of resilient plant species while utilizing nutrient replenishment techniques to restore soil organic matter.

Fluctuations in hydrologic patterns

- Avoid constructing excessively steep slopes to prevent erosion and land instability.
- Ensure that excavations and earthworks in ecologically sensitive areas are monitored by a qualified palaeontologist to protect valuable environmental and paleontological resources.

Environmental air degradation

- Minimize the extent of vegetation removal to preserve the natural landscape.
- Reduce on-site vehicle usage and implement regular preventative maintenance to ensure optimal combustion efficiency and minimal emissions.
- Develop a site-specific dust control plan to mitigate air quality impacts.
- Construct wind fences around disturbed areas to prevent dust and debris from affecting surrounding environments beyond the site boundaries.
- Inspect and clean equipment vehicle tires before accessing paved public roads to prevent dirt tracking and contamination.
- Comply with state emission regulations for all combustion sources to reduce environmental pollution.

Reduce carbon emissions Positive impact.

Composite Materials: SEAJCCM2024
Materials Research Proceedings 56 (2025) 53-68

Materials Research Forum LLC
https://doi.org/10.21741/9781644903636-6

OCCUPATIONAL SAFETY AND HEALTH IMPACT

Unsafe workplace conditions	• Establish regulations for facility operations to ensure the safe and compatible use of materials. • If pesticides or herbicides are necessary, develop a comprehensive plant management plan to minimize environmental impact. • Conduct thorough assessments of previously utilized areas to identify and mitigate potential hazardous material contamination.
Adverse health effects on workers and nearby areas	When conducting health risk assessments, it is important to consider the potential exposure of workers to both non-carcinogenic and carcinogenic hazards during the development and operation of the facility.
Hazardous material endangers to public health	Design and operate systems that handle hazardous materials in a manner that minimizes the risk of their release.
Production of construction-related solid waste	• Compile a comprehensive inventory of all hazardous materials to be used, stored, transported, or disposed of during the project. • Develop a waste management plan that outlines expected solid and liquid waste streams, including procedures for waste classification, inspection, reduction, and appropriate storage, as well as the management and disposal requirements for each waste type. • Ensure that surplus structural steel components, such as metal poles and rods, are not transported, stored, or lifted in areas where they could meet overhead power lines.
Heightened danger of fires and environmental pollution resulting from improper material storage	• Develop a fire protection and prevention plan to minimize the risk of fires caused by chemicals and materials stored on-site, particularly heat exchange fluids that are highly flammable. • Ensure compliance with all required protocols when storing and transporting explosive materials and blasting equipment. • At every project stage, implement preventive measures against wildfires by creating and executing appropriate wildfire control strategies. These measures should include training for workers, as well as ongoing inspection and monitoring procedures.

Conclusion

Environmental considerations play a critical role in the development of large-scale solar photovoltaic (LSSPV) systems in Pulau Pinang, Malaysia. This study employed Environmental Impact Screening (EIS) as the preliminary phase of the Environmental Impact Assessment (EIA) to systematically assess the potential environmental impacts associated with LSSPV projects. The research successfully fulfilled its key objectives by identifying potential environmental harms arising from solar farming and collecting valuable data through site analysis and expert consultations. Using the Analytic Hierarchy Process (AHP), the study categorized the impacts into three primary areas:

- Economic impact
- Ecological and chemical impact
- Occupational safety and health (OSH) impact

The AHP methodology was utilized to systematically analyses the data gathered, while mitigation strategies were proposed through the Multi-Criteria Decision-Making (MCDM) approach. These strategies are imperative for fostering sustainable development while mitigating environmental risks. This research underscores the pivotal role of strategic foresight and comprehensive impact assessments in the successful deployment of large-scale solar projects. By adopting sound planning and effective mitigation measures, it is possible to attenuate, if not wholly eliminate, adverse environmental consequences. As societies transition toward a decarbonized future, the integration of solar energy with advanced energy storage solutions and carbon capture technologies will be crucial for advancing sustainability. This integrative approach will not only expedite the proliferation of solar energy but will also fortify the importance of environmental impact assessments on both national and global fronts.

Considering the intricate nature of environmental challenges, efficient optimization, harmonized policies, and ongoing research are essential to maximize both socioeconomic and environmental advantages. Moreover, this study serves as a crucial reference for future large-scale solar projects, offering valuable insights to key stakeholders—such as government authorities, developers, and regulatory bodies—on effective mitigation strategies to minimize environmental risks while enhancing positive outcomes. Adherence to environmental sustainability objectives will not only optimize economic, ecological, and chemical benefits but also strengthen occupational safety and health practices, ultimately mitigating environmental degradation in Pulau Burung, Pulau Pinang.

Nonetheless, the study is subject to certain limitations. The findings are primarily based on input from experts and authorities, which may not fully capture the perspectives of local communities. Future research should integrate data from local stakeholders to refine the methodology and enhance the accuracy of environmental impact assessments. By addressing these gaps, subsequent studies will be able to adopt a more holistic approach to sustainable solar energy development.

References

[1] K.E. Sarah, U. Roland, E.O. Ephraim, A review of solar photovoltaic technologies, International Journal of Engineering Research & Technology 9 (2020) 741-749. https://doi.org/10.17577/IJERTV9IS070244

[2] L.Y. Seng, G. Lalchand, G.M.S. Lin, Economical, environmental and technical analysis of building integrated photovoltaic systems in Malaysia, Energy policy 36 (2008) 2130-2142. https://doi.org/10.1016/J.ENPOL.2008.02.016

[3] M. Almaktar, H. Abdul Rahman, M.Y. Hassan, W.Z. Wan Omar, Photovoltaic technology in Malaysia: past, present, and future plan, International Journal of Sustainable Energy 34 (2015) 128-140. https://doi.org/10.1080/14786451.2013.852198

[4] M. Zarzavilla, A. Quintero, M.A. Abellán, F.L. Serrano, M.C. Austin, N. Tejedor-Flores, Comparison of environmental impact assessment methods in the assembly and operation of photovoltaic power plants: A systematic review in the castilla—La mancha region, Energies 15 (2022) 1926. https://doi.org/10.3390/en15051926

[5] A. Quek, A.R. Abbas, N.Z.I.S. Zaman, W.N.S.W. Ata, M.F. Zainal, F.K.M. Yapandi, Z.F. Ibrahim, A. Suhardi, Life Cycle Assessment of Large-Scale Solar Photovoltaic Plant based in

Malaysia, IOP Conference Series: Earth and Environmental Science 1372 (2024) 012055.
https://doi.org/10.1088/1755-1315/1372/1/012055

[6] N.I. Mustafa, N.A. Ludin, N.M. Mohamed, M.A. Ibrahim, M.A.M. Teridi, S. Sepeai, A. Zaharim, K. Sopian, Environmental performance of window-integrated systems using dye-sensitised solar module technology in Malaysia, Solar Energy 187 (2019) 379-392.
https://doi.org/10.1016/J.SOLENER.2019.05.059

[7] M. Tawalbeh, A. Al-Othman, F. Kafiah, E. Abdelsalam, F. Almomani, M. Alkasrawi, Environmental impacts of solar photovoltaic systems: A critical review of recent progress and future outlook, The Science of the total environment (2020) 143528.
https://doi.org/10.1016/j.scitotenv.2020.143528

[8] M. Sahid, R. Suratman, H. Ali, Determination of Special Permit Rate For Largescale Solar Development In Johor Based On Planner Perspective, Planning Malaysia 19 (2021) 1-11.
https://doi.org/10.21837/pm.v19i18.1027

[9] K.A.C. Kamaruddin, N.N. Mansor, N.A.M.I. Yeong, A.F. Othman, Optimal Integration of Large-Scale Solar and Battery Energy Storage System in Malaysia, 2023 Innovations in Power and Advanced Computing Technologies (i-PACT) (2023) 1-7. https://doi.org/10.1109/i-PACT58649.2023.10434585

[10] M.A.M. Khan, Y.I. Go, Design, optimization and safety assessment of energy storage: A case study of large-scale solar in Malaysia, Energy Storage 3 (2021) e221.
https://doi.org/10.1002/est2.221

[11] M. Vaka, R. Walvekar, A.K. Rasheed, M. Khalid, A review on Malaysia's solar energy pathway towards carbon-neutral Malaysia beyond Covid'19 pandemic, Journal of cleaner production 273 (2020) 122834. https://doi.org/10.1016/j.jclepro.2020.122834

[12] I. Jamil, H. Lucheng, S. Habib, M. Aurangzeb, A. Ali, E.M. Ahmed, Performance ratio analysis based on energy production for large-scale solar plant, IEEE Access 10 (2022) 5715-5735. https://doi.org/10.1109/ACCESS.2022.3141755

[13] M. Mohanan, Y.I. Go, Optimized power system management scheme for LSS PV grid integration in Malaysia using reactive power compensation technique, Global Challenges 4 (2020) 1900093. https://doi.org/10.1002/gch2.201900093

[14] A. Morrison-Saunders, J. Bailey, Exploring the EIA/environmental management relationship, Environmental Management 24 (1999) 281-295.
https://doi.org/10.1007/S002679900233

[15] J.J. Loomis, M. Dziedzic, Evaluating EIA systems' effectiveness: a state of the art, Environmental Impact Assessment Review 68 (2018) 29-37.
https://doi.org/10.1016/J.EIAR.2017.10.005

[16] A. Quek, A.R. Abbas, N.Z.I.S. Zaman, W.N.S.W. Ata, M.F. Zainal, F.K.M. Yapandi, Z.F. Ibrahim, A. Suhardi, Life Cycle Assessment of Large-Scale Solar Photovoltaic Plant based in Malaysia, IOP Conference Series: Earth and Environmental Science 1372 (2024) 012055.
https://doi.org/10.1088/1755-1315/1372/1/012055

[17] H. Taha, The potential for air-temperature impact from large-scale deployment of solar photovoltaic arrays in urban areas, Solar Energy 91 (2013) 358-367.
https://doi.org/10.1016/J.SOLENER.2012.09.014

Composite Materials: SEAJCCM2024
Materials Research Proceedings 56 (2025) 53-68

Materials Research Forum LLC
https://doi.org/10.21741/9781644903636-6

[18] M.K.H. Rabaia, M.A. Abdelkareem, E.T. Sayed, K. Elsaid, K.J. Chae, T. Wilberforce, A.G. Olabi, Environmental impacts of solar energy systems: A review, Science of The Total Environment 754 (2021) 141989. https://doi.org/10.1016/j.scitotenv.2020.141989

[19] O.A. Agboola, C.T. Downs, G. O'Brien, Ecological risk of water resource use to the wellbeing of macroinvertebrate communities in the rivers of KwaZulu-Natal, South Africa, Frontiers in Water 2 (2020) 584936. https://doi.org/10.3389/frwa.2020.584936

[20] B.L. Golden, E.A. Wasil, P.T. Harker, The analytic hierarchy process, Applications and Studies 2 (1989) 1-273. https://doi.org/10.1007/978-3-642-50244-6

[21] T.L. Saaty, K.P. Kearns, Chapter 3—The Analytic Hierarchy Process, Analytical planning (1985) 19-62. https://doi.org/10.1016/B978-0-08-032599-6.50008-8

[22] E. Cheng, H. Li, Analytic hierarchy process, Measuring Business Excellence 5 (2001) 30-37. https://doi.org/10.1108/EUM0000000005864

[23] H.K. Chan, X. Sun, S.H. Chung, When should fuzzy analytic hierarchy process be used instead of analytic hierarchy process?, Decision Support Systems 125 (2019) 113114. https://doi.org/10.1016/J.DSS.2019.113114

[24] W. Ho, Integrated analytic hierarchy process and its applications–A literature review, European Journal of operational research 186 (2008) 211-228. https://doi.org/10.1016/j.ejor.2007.01.004

[25] Y. Zhang, Z. Tian, B. Liu, S. Chen, J. Wu, Effects of photovoltaic power station construction on terrestrial ecosystems: A meta-analysis, Frontiers in Ecology and Evolution 11 (2023) 1151182. https://doi.org/10.3389/fevo.2023.1151182

[26] R.D. Simkin, K.C. Seto, R.I. McDonald, W. Jetz, Biodiversity impacts and conservation implications of urban land expansion projected to 2050, Proceedings of the National Academy of Sciences 119 (2022) e2117297119. https://doi.org/10.1073/pnas.2117297119

[27] D.E. El Badry Fadl, The role of environmental impact assessment in achieving sustainable development, International Journal of Modern Agriculture and Environment 1 (2021) 38-50. https://doi.org/10.21608/ijmae.2023.214940.1008

Composite Materials: SEAJCCM2024
Materials Research Proceedings 56 (2025) 69-78

Materials Research Forum LLC
https://doi.org/10.21741/9781644903636-7

Synthesis and Characterization of Modified Nanocellulose-Based Adsorbent for Removal of Cu²⁺ Ion Metal in Wastewater

Noorhaslin CHE SU[1,a], Ain Aqilah BASIRUN[1,b],
Nor Shahroon HAMEED SULTAN[1,c], Devagi KANAKARAJU[2,d] and
Cecilia Devi WILFRED[1,3,e] *

[1]Centre of Research in Ionic Liquid, Universiti Teknologi PETRONAS, Persiaran UTP, 32610 Seri Iskandar, Perak, Malaysia

[2]Faculty of Resource Science and Technology, Universiti Malaysia Sarawak, 94300 Kota Samarahan, Sarawak, Malaysia

[3]Fundamental and Applied Sciences Department, Universiti Teknologi PETRONAS, Persiaran UTP, 32610 Seri Iskandar, Perak, Malaysia

[a]haslincs@gmail.com, [b]ainaqilahbasirun@gmail.com, [c]norshahroon23@gmail.com, [d]kdevagi@unimas.my, [e]cecili@utp.edu.my

Keywords: Nanocellulose, Cu (II), Magnetite, TiO₂

Abstract. The continuous release of heavy metal from various industrial activities has been recognized as one of the major causes contributing to water pollutant. Herein, a novel magnetite adsorbent consisting of nanocellulose (NC), magnetite nanoparticles (MNP) and titanium dioxide (TiO₂) have been prepared using hydrothermal method. The adsorbent material with adsorption, magnetic and photocatalytic properties has been successfully synthesized and used for Cu²⁺ removal from industrial wastewater. The magnetite adsorbent made up of nanocellulose from sago was characterized by X-ray diffraction (XRD), Fourier Transform Infrared Spectroscopy, (FTIR), field emission scanning electron microscope (FESEM) for physicochemical study. XRD demonstrating high crystallinity of nanocellulose and displays well crystallized anatase TiO₂. TiO₂ embedded on the surface on NC/MNP were clearly observed through FESEM. The photocatalytic efficiency of the adsorbent degradation of Cu was evaluated using atomic adsorption spectroscopy (AAS). The NC/MNP showed removal of 54.1% of Cu²⁺ metal. Increase of TiO₂ loading (800µl) facilitate Cu removal up to 96% after 120 minutes exposure under ultraviolet irradiation. Kinetic analysis using the PSO model at 100mg/L of Cu²⁺ gave a value of equilibrium adsorption capacity, q_e of 4.29 mg/g and a value of the pseudo-second-order rate constant, K_2 of 0.004 (95% C.I., 0.0037 to 0.012). The result from these studies showed that the combination of nanocellulose, magnetite and TiO₂ has potential heavy metal removals.

Introduction

Rapid urbanization, industrialization and growing population have led a serious water pollution. Discharging untreated heavy metals such as Hg^{2+}, Cu^{2+}, Ni^{2+}, and Cr^{7+} into river or sea can give band effect to human health and disturb the aquatic system. Cu^{2+} is one of the common heavy metals in wastewater originated from the effluents of electroplating, electrical industry, wire mills and other industrial plants. Accumulation of Cu^{2+} in human body above 1.3mg/L level enriched through the food chain for short period of time ions can cause mucosal irritation, intestinal problems widespread capillary and renal damage [1].

Nanocellulose (NC) based materials in recent years, which can offer promising results in wastewater purification [2]. The prospect of developing cellulose into nanocellulose by different physicochemical approaches has opened a new field for research into the advantages and possibilities of wastewater treatment. The enormous hydroxyl group (-OH) in nanocellulose can

Content from this work may be used under the terms of the Creative Commons Attribution 3.0 license. Any further distribution of this work must maintain attribution to the author(s) and the title of the work, journal citation and DOI. Published under license by Materials Research Forum LLC.

Composite Materials: SEAJCCM2024
Materials Research Proceedings 56 (2025) 69-78

Materials Research Forum LLC
https://doi.org/10.21741/9781644903636-7

efficiently anchor metal cations for heavy metal removal [3-5]. To effectively remove pollutants, NC has a remarkable explicit mechanical behaviour and chemically changed surface functioning that favoured the waste absorption [6]. On top of that, NC can be more cost-effective than other nanomaterials like graphene and ceramics owing to their low cost of the raw material and simple processing.

Several approaches have been reported for removing organic impurities from wastewater including the adsorption technique [7], membrane separation [8] and advanced oxidation process (AOP) [9] etc. Adsorption is the process by which a pollutant, such as organic compound and heavy metal ions are immobilized to the surface of a solid object by chemical or electrostatic in water treatment activity [10]. Magnetite (Fe_3O_4) nanoparticles have attracted scientific attention to be used as adsorbent due to their outstanding performance, such as their ability to display superparamagnetic properties with high saturation magnetization in the presence of external magnetic field [11]. Magnetite served with high surface area, good compatibility, low toxicity, and good physicochemical stability have been developed for potential application to detoxification of groundwater, surface water and industrial effluents in wastewater treatment [12].

Photocatalysis as part of the AOP process been extensively studied. Generally, photocatalysis consists of semiconductor structures. The most common semiconductor used for photocatalysis is TiO_2. As the TiO_2 semiconductor is illuminates by light, the TiO_2 gain equal or greater photon energy between VB and CB, the electron (e^-) from valance band (VB) jump to conduction band (CB) and leave positively charged hole (h^+) on VB. This pair of e^- and h^+ can respectively migrate to the surface of the semiconductor to undergo series of oxidation and reduction which are embodied in the conversion of different valance states in the treatment of heavy metal. Photocatalysis also known as a "catalytic" reaction by adsorption of light which could mineralize organic compound into non-toxic product such as carbon dioxide and water [13] [14-16]. The possible mechanism occur during TiO_2 photocatalyst are shown in Eq. 1 [17].

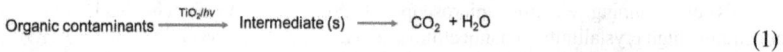

$$\text{Organic contaminants} \xrightarrow{TiO_2/h\nu} \text{Intermediate (s)} \longrightarrow CO_2 + H_2O \tag{1}$$

This study will evaluate an adsorbent consisting of nanocellulose from sago, Fe_3O_4 and TiO_2 for synergistic interaction of adsorption and photocatalysis in removing Cu^{2+} from wastewater.

Materials and Method

Chemicals

Nanocellulose was obtained from sago bark and sago shell samples collected from Sarawak. Iron (III) hexahydrate ($FeCl_3.6H_2O$), iron (II) chloride tetrahydrate ($FeCl_2.4H_2O$), trisodium citrate ($Na_3C_6H_5O_7$) and sodium hydroxide (NaOH) from Merck (Germany). Titanium (IV) n-butoxide (TBOT) and ethanol were purchased from Merck.

Instrumentation

The powder phase compositions were identified with X-ray diffraction (XRD) equipment (D/max 2550 PC, Rigaku Co., Japan) using Cu Kα radiation at 40Kv and 200mA. The functional groups of the nanocellulose adsorbents were determined using Fourier-transform infrared spectrophotometer (FTIR) instrument (Model Frontier, Thermo Fisher Scientific, Waltham, MA, USA). The morphology was observed by Field Emission Scanning Electron Microscope (FESEM) (Supra 55VP Carl Zeiss, Germany) equipped with an energy dispersive X-ray and field emission scanning electron microscope (Clara Tescan). The nitrogen adsorption and desorption isotherm were obtained at 77K using Autosorb -1 MP (Quantachrome, USA) utilizing Barret-Emmett-Teller calculations of pore volume and pore size (diameter) distributions from the desorption branch of

Composite Materials: SEAJCCM2024   Materials Research Forum LLC
Materials Research Proceedings 56 (2025) 69-78   https://doi.org/10.21741/9781644903636-7

the isotherm. The concentration of Cu^{2+} ions was determined by an atomic absorption spectrometer (AAS) from Agilent (Model 240 FS, Agilent Technologies, United States). The non-linear regression of all kinetic models was plot and analysed using Curve Expert 6.0.

Preparation of nanocellulose

1 g of was extracted cellulose from sago was hydrolysed with 20 mL of acid and H_2SO_4 (40 wt.%) with ratio 1:20 at 45 °C for 60 min. The suspension was washed repeatedly using cold deionized water (centrifuge at 4000 rpm, 30 min) to remove the excess H_2SO_4 until pearl white with a neutral pH sedimentation were obtained. The slurry was further treated by sonication for 40 min in a sonicator before placing in a freezer-dryer. The nanocellulose powder was characterized by X-ray diffraction (XRD), infrared (FTIR, and field emission scanning electron microscopy (FESEM) analysis.

Preparation of nanocellulose/Fe₃O₄

0.5 g nanocellulose was dissolved in 20ml of H_2O with vigorous mechanical stirring at 80 °C in a water bath until a homogenous solution was formed. Subsequently, 0.45 g Fe_3O_4 and 1.21 g of sodium acetate were added in a viscous solution of nanocellulose with vigorous agitating for 30 min at 80 °C in a water bath to form a homogeneous solution. Then, the solution was poured into a Teflon-lined stainless-steel autoclave for hydrothermal. Later cooling to ambient temperature, the black precipitate product was separated by a magnet and washed several times with water and ethanol. The as-prepared sample was dried overnight at 50 °C in the oven.

Preparation of nanocellulose/ Fe₃O₄/TiO₂

0.10 g nanocellulose/Fe_3O_4 powder was added into 10 mL absolute ethanol solution and sonicated for ca. 20 min in an ultrasonic bath. 200 μL of TBOT was added into the dispersion dropwise, and the mixture was sonicated for 15 min. It was then decanted into a vapor phase hydrolysis apparatus (VPHA) with deionized water situated at the bottom of it to produce vapor at raised temperature. The closed VPHA was heated to 110°C and maintained for 5 hr. Then, the VPHA was cooled to room temperature. The as-prepared powder was collected by magnet and washed with absolute ethanol and deionized water and dried in the oven at 50 °C. Table 1 describes the two types of adsorbent prepared.

Table 1. Prepared adsorbent.

Adsorbent	Sample code
Nanocellulose/Fe_3O_4	Ads-1
Nanocellulose/Fe_3O_4/TiO_2	Ads-2

Photocatalytic reaction

The photocatalytic activity of adsorbent Ads-1 and Ads-2 was evaluated by activity of the adsorbent in terms of removal of Cu^{2+} in wastewater under UV-light illumination using 100-W UV lamp (UVP Co.Upland),Canada) at 365nm. Initial concentration of Cu^{2+} was 100mg/L. In the experiment, the suspension was stirred magnetically (200 rpm) at dark for 30 min. After 30 minutes in dark, 10 ml sample was taken from the wastewater by means of syringe filter. This method was repeated for 30, 60, 90 and 120 min under UV-light illumination. Collected samples were analysed with AAS to determine the concentration of Cu^{2+}. The Cu^{2+} removal was calculated by the following formula from the following Eq. 2 [12, 18].

$$R\ (\%) = [C_0 - C_1)/C_0] \times 100\ \% \tag{2}$$

Where R is the removal percentage of Cu^{2+}; and C_0 is the initial concentration of Cu^{2+} (mg/L); C_1 is the concentration of Cu^{2+} at reaction time (mg/L).

Composite Materials: SEAJCCM2024   Materials Research Forum LLC
Materials Research Proceedings 56 (2025) 69-78   https://doi.org/10.21741/9781644903636-7

Results and Discussion

Physicochemical Study

Fig. 1 shows XRD peak for nanocellulose, Ads-1 and Ads-2. The nanocellulose shows the appearance peak at 15.28 and 22.94 [19]. The Ads-1 shows the appearance of the peak 22.94°, 30.06°, 35.76°, 43.5°, 57.11° and 62.5°. The peaks of nanocellulose 15.28° disappear with addition of Fe_2O_3 and the intensity of the nanocellulose peak at 22.94° were less for both Ads-1 and Ads-2 which, corresponds to the interaction of nanocellulose and magnetite. The dominate peaks were observed in both Ads-1 and Ads-2 at 30.06°, 35.76°, 43.5°, 57.11° and 62.5° which, represents the present of Fe_3O_4 [11]. The Ads-2 shows similar peaks with Ads-1, with additional peaks observed at 25.26° and 54.3°. The peaks represent the anatase phase TiO_2 peak [20] which, demonstrates the formation of well-crystallized TiO_2.

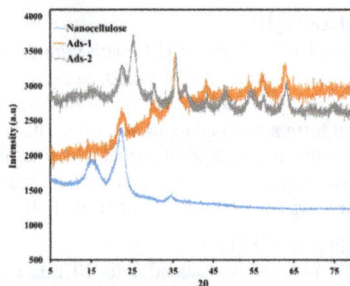

Fig. 1. XRD pattern of nanocellulose, Ads-1 and Ads-2.

Fig. 2 shows the FTIR spectra of the nanocellulose, Ads-1 and Ads-2. The wide absorption band at 3403 cm^{-1} and 1645cm^{-1} can be attributed to the stretching vibration of O-H from the hydroxyl group [21-22]. The absorption band at 2901 cm^{-1} is associated with –C-H stretching vibrations [23]. The characteristic cellulose peak of pyranose ring was noticed at 1063 cm^{-1}[4]. Meanwhile, the high intensity band for iron oxide at 589 cm^{-1} was found in the prepared Ads-1and Ads-2 [22, 24]. This peak confirms the successful formation of Fe_3O_4 nanoparticles on the surface of nanocellulose.

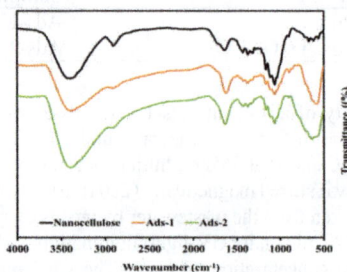

Fig. 2. FTIR spectra of nanocellulose, Ads-1 and Ads-2.

Fig. 3(a-d) shows the FESEM images of raw sago, treated sago with KOH and acid hydrolysis and the prepared adsorbent, respectively. From the figures, the morphology of the sago changes with each treatment. The smooth surface of raw sago is shown as in Fig. 3(a). The presence of a smooth surface represents the hemicellulose and lignin in the amorphous state [25]. The fibre

Composite Materials: SEAJCCM2024

Materials Research Proceedings 56 (2025) 69-78

Materials Research Forum LLC

https://doi.org/10.21741/9781644903636-7

surface becomes rougher after being treated with KOH (Fig. 3(b)). This could indicate the partial removal of the outer non-cellulosic layer such as hemicelluloses, lignin, pectin, wax, and other impurities contained in sago [26-27]. Fig. 3(c) depicts the nanocellulose after acid hydrolysis. The acid hydrolysis breaks the fibres into nanoscale [28]. Fig. 3(d) represents the prepared adsorbent. White particle on Ads-2 surface was identified as TiO_2 particles. This finding shows that TiO_2 particles were well dispersed on the nanocellulose and magnetite surface.

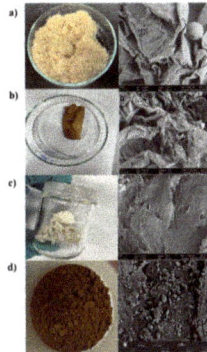

Fig. 3. FESEM images of (a) raw sago, (b) treated sago with KOH, (c) nanocellulose, and (d) prepared adsorbent.

The N_2 adsorption –desorption isotherm shown in Fig. 4 of adsorbent Ads-1 and Ads-2 presents a type-IV like curve with H3 hysteresis loop, characteristics of mesopores in the spherical cores and shells. The hysteresis loop of both Ads-1 and Ads-2 illustrates the fact that the larger crystallite stacking larger intraparticle pores at elevated temperature. Type IV isotherms show the cooperative adsorption where the binding site on the surface is identical, which can host multiple molecules. This applies to this present study since the adsorbent consists of nanocellulose and Fe_3O_4 components which provide high affinity to various adsorbates [29].

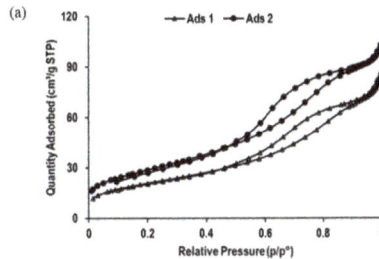

Fig. 4. N_2 Adsorption –desorption isotherm of Ads -1 and Ads-2.

Table 2 details the pore area and diameter of both prepared adsorbents. The average pore size of Ads-1 and Ads-2 are 6.853 nm and 6.011 nm, respectively. The pore diameter shows a decrement in value due to the agglomeration of TiO_2 which accumulates the surface of the adsorbent. The pore diameter of both samples show that adsorbent surface are mesopores. This data agrees with previous study which confirms that KOH modification of nanocellulose improves the porosity and enhance the affinity of the adsorbent towards pollutants.

Composite Materials: SEAJCCM2024
Materials Research Proceedings 56 (2025) 69-78

Materials Research Forum LLC
https://doi.org/10.21741/9781644903636-7

Table 2. The results of BET analyses for the prepared adsorbents.

Adsorbent	BET surface area (m²/g)	Pore diameter (nm)
Ads-1	74.87	6.853
Ads-2	103.39	6.011

Removal Studies of Cu²⁺ ion in wastewater

Fig. 5 compares the removal percentage of the prepared adsorbent. Ads-1 and Ads-2 showed Cu^{2+} removal of 54.1% and 96% respectively. The presence of nanocellulose and Fe_3O_4 in Ads-1 enables it to work as adsorbent to remove Cu^{2+} in wastewater. It is known that material with large amount amounts of active groups such as hydroxyl or carboxyl group in nanocellulose can absorb contaminants from wastewater [31]. As for Ads 2, there is a large improvement when TiO_2 is added to the adsorbent. TiO_2 has the ability of harvest light energy to stimulate the chemical reaction within the system. Heavy metal Cu^{2+} is removed almost completely from the wastewater. Table 3 displays previous studies where nanocellulose is used as adsorbent for heavy metal removal. Our work is comparable with the earlier work and shows that sago based nanocellulose + Fe_3O_4 + TiO_2 is a promising adsorbent in the near future.

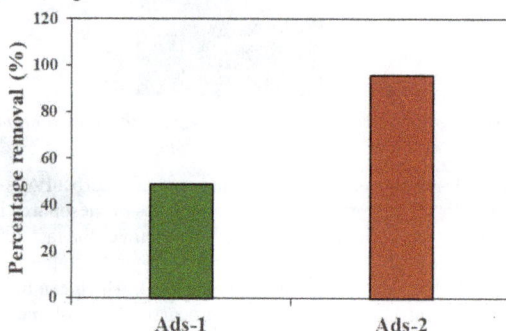

Fig. 5. Percentage of Cu^{2+} removal for Ads-1 and Ads-2.

Table 3. Comparative study of Cu2+ ion removal by different nanocellulose sources.

Nanocellulose sources	Metal ions (removal capacity)	References
Sago based NC + Fe_3O_4 + TiO_2	Cu^{2+} (96 %)	This study
TEMPO-oxidized cellulose nanoparticles (TOCN)	Cu^{2+} (75 mg/g)	[32]
Oryza sativa husk-NC	Cu^{2+} (91 %)	[33]
Chitosan-NC	Cu^{2+} (98 %)	[34]

Kinetic study

Kinetic models can provide information on adsorption pathways and the mechanisms that may be involved. This is also significant information for the process development and adsorption system design. Kinetics of 100 ppm Cu^{2+} removal using photocatalysis reaction is modelled using (a) pseudo–first order and (b) pseudo-second order a shown in Fig 6. The list of other error functions analysis is presented in Table 4 with pseudo–second order model giving the best values with the lowest AICc, RSME values of 0.56, 0.53, respectively and highest correlation coefficient, R^2 of 0.99, with bias and accuracy factors near unity. The pseudo–second order theory was proposed by Blanchard et al. [35] and the derivation was developed by Azizian [36] assuming the initial adsorbate concentration is not too high as compared to the term derivation equation developed. Thus, the adsorption process obeyed the pseudo–second at lower initial Cu^{2+} concentration. In the pseudo-second order reaction, the rate-controlling step is linked to chemisorption. In this case, the adsorption mechanism is controlled chemically by the rate of adsorption. At low sorbate/sorbent ratios (first order at extremely low ratios), the sorption kinetics are often governed by a reversible second-order reaction, and at larger sorbate/sorbent ratios, the sorption kinetics are governed by two reversible second-order reactions that are competitive with one another. Kinetic analysis using the PSO model at 100mg/L of Cu^{2+} gave a value of equilibrium adsorption capacity, q_e of 4.29 mg/g and a value of the pseudo-second-order rate constant, K_2 of 0.004 (95% C.I., 0.0037 to 0.012) (Table 5).

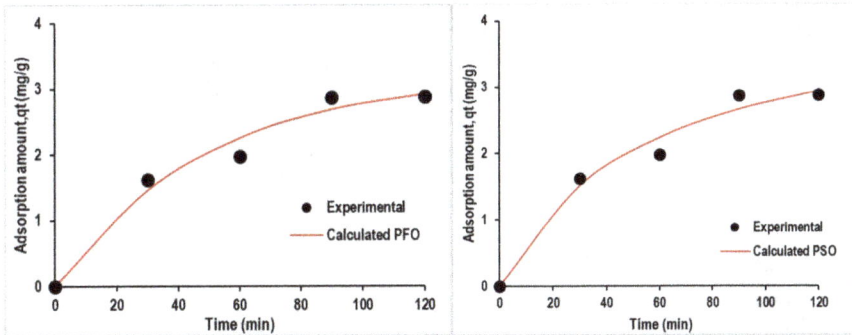

Fig. 6. Kinetics of 100 ppm Cu^{2+} removal using photocatalysis reaction as modelled using the (a) pseudo–first order and (b) pseudo-second order.

Table 4. Error function analysis of selective kinetics models.

Model	Parameter	SSE	R_2	AICc	AF	BF
Pseudo-1st order	2	0.229	0.990	-13.974	1.088	0.9994
Pseudo-2nd order	2	0.223	0.991	-14.243	1.034	0.9970

Table 5. Calculated maximum adsorption capacity of selective kinetic models and constant k for Cu^{2+} ion adsorption.

Model	Parameter	Maximum adsorption capacity, q_e (mg/g)	constant, k (95% confidence interval)
Pseudo-1st order	2	3.199	0.04 (0.023 to 0.05)
Pseudo-2nd order	2	4.291	0.004(0.0037 to 0.012)

Summary

In summary, this research has demonstrated that the synergistic interaction of adsorption and photocatalysis reaction for environmental purification is newly established and could be considered as one of the most efficient techniques. The result from these studies showed that the combination of nanocellulose, magnetite and TiO_2 has potential for heavy metal removals.

References

[1] Q. Wu, N.F. Tam, J.Y. Leung, X. Zhou, J. Fu, B. Yao, L. Huang, Ecological risk and pollution history of heavy metals in Nansha mangrove, South China, Ecotoxicology and Environmental Safety 104 (2014) 143-151. https://doi. 10.1016/j.ecoenv.2014.02.017

[2] Y. Ahn, B.E. Logan, Effectiveness of domestic wastewater treatment using microbial fuel cells at ambient and mesophilic temperatures, Bioresource Technology 101 (2010) 469-475. https://doi.10.1016/j.biortech.2009.07.039

[3] C.N.C. Hitam, A.A. Jalil, Recent advances on nanocellulose biomaterials for environmental health photoremediation: An overview, Environmental Research 204 (2022) 111964. https://doi.10.1016/j.envres.2021.111964

[4] L.T. Yogarathinam, P.S. Goh, A.F. Ismail, A. Gangasalam, N.A. Ahmad, A.Samavati, S.C. Mamah, M.N.Z. Abidin, N.C. Ng, B. Gopal, Nanocrystalline cellulose incorporated biopolymer tailored polyethersulfone mixed matrix membranes for efficient treatment of produced water, Chemosphere 293 (2022) 133561. https://doi.10.1016/j.chemosphere.2022.133561

[5] F. Asempour, D. Emadzadeh, T. Matsuura, B. Kruczek, Synthesis and characterization of novel Cellulose Nanocrystals-based Thin Film Nanocomposite membranes for reverse osmosis applications, Desalination 439 (2018) 179-187. https://doi. 10.1016/j.desal.2018.04.009

[6] T.K. Das, A. Poater, Review on the use of heavy metal deposits from water treatment waste towards catalytic chemical syntheses, International Journal of Molecular Sciences 22 (2021) 13383. https://doi.0.3390/ijms222413383

[7] G. Lofrano, M. Carotenuto, G. Libralato, R.F. Domingos, A. Markus, L. Dini, R.K. Gautam, D. Baldantoni, M. Rossi, S.K. Sharma, M.C. Chattopadhyaya, M. Giugni, S. Meric, S. Polymer functionalized nanocomposites for metals removal from water and wastewater: an overview, Water Research 92 (2016) 22-37. https://doi.10.1016/j.watres.2016.01.033

[8] M.S. Denny Jr, S.M. Cohen, *In situ* modification of metal–organic frameworks in mixed-matrix membranes, Angewandte Chemie International Edition 54 (2015) 9029-9032. https://doi.10.1002/anie.201504077

[9] D. Kanakaraju, Y.C. Lim, A. Pace, Magnetic hybrid $TiO_2/Alg/FeNPs$ triads for the efficient removal of methylene blue from water, Sustainable Chemistry and Pharmacy 8 (2018) 50-62. https://doi.10.1016/j.scp.2018.02.001

[10] A. Mautner, Nanocellulose water treatment membranes and filters: a review, Polymer International 69 (2020) 741-751. https://doi.10.1002/pi.5993

[11] J. Chang, Q. Zhang, Y. Liu, Y. Shi, Z. Qin, Preparation of Fe_3O_4/TiO_2 magnetic photocatalyst for photocatalytic degradation of phenol, Journal of Materials Science: Materials in Electronics 29 (2018) 8258-8266. https://doi. 10.1007/s10854-018-8832-7

[12] H.S. Hassan, M.F. Elkady, A.A. Farghali, A.M. Salem, A.I. Abd El-Hamid, Fabrication of novel magnetic zinc oxide cellulose acetate hybrid nano-fiber to be utilized for phenol decontamination, Journal of the Taiwan Institute of Chemical Engineers 78 (2017) 307-316. https://doi.10.1016/j.jtice.2017.06.021

[13] R.L. Narayana, M. Matheswaran, A. Abd Aziz, Saravanan, P. Photocatalytic decolourization of basic green dye by pure and Fe, Co doped TiO_2 under daylight illumination, Desalination 269 (2011) 249-253. https://doi.10.1016/j.desal.2010.11.007

[14] Y. Ku, I.L. Jung, Photocatalytic reduction of Cr (VI) in aqueous solutions by UV irradiation with the presence of titanium dioxide, Water Research 35 (2001) 135-142. https://doi.10.1016/S0043-1354(00)00098-1

[15] M.J. López-Muñoz, J. Aguado, A. Arencibia, R. Pascual, Mercury removal from aqueous solutions of $HgCl_2$ by heterogeneous photocatalysis with TiO_2, Applied Catalysis B: Environmental 104 (2011) 220-228. https://doi. 10.1016/j.apcatb.2011.03.029

[16] K.M. Joshi, B.N. Patil, D.S. Shirsath, V.S. Shrivastava, Photocatalytic removal of Ni (II) and Cu (II) by using different Semiconducting materials, Advances in Applied Science Research 2 (2011) 445-54.

[17] M.N. Chong, B. Jin, C.W. Chow, C. Saint, Recent developments in photocatalytic water treatment technology: a review, Water Research 44 (2010) 2997-3027. https://doi.10.1016/j.watres.2010.02.039

[18] Q. Sun, H. Li, B. Niu, X. Hu, C. Xu, S. Zheng, Nano-TiO_2 immobilized on diatomite: characterization and photocatalytic reactivity for Cu^{2+} removal from aqueous solution, Procedia Engineering 102 (2015) 1935-1943. https://doi.10.1016/j.proeng.2015.01.334

[19] S. Naduparambath, E. Purushothaman, Sago seed shell: determination of the composition and isolation of microcrystalline cellulose (MCC), Cellulose 23 (2016) 1803-1812. https://doi.10.1007/s10570-016-0904-3

[20] H. Zhang, X. He, W. Zhao, Y. Peng, D. Sun, H. Li, X. Wang, Preparation of Fe_3O_4/TiO_2 magnetic mesoporous composites for photocatalytic degradation of organic pollutants, Water Science and Technology 75 (2017) 1523-1528. https://doi.10.2166/wst.2017.002

[21] S.B. Hammouda, N. Adhoum, L. Monser, Synthesis of magnetic alginate beads based on Fe_3O_4 nanoparticles for the removal of 3-methylindole from aqueous solution using Fenton process, Journal of Hazardous Materials 294 (2015) 128-136. https://doi.10.1016/j.jhazmat.2015.03.068

[22] N. Amiralian, M. Mustapic, M.S.A. Hossain, C. Wang, M. Konarova, J. Tang, J. Na, A. Khan, A. Rowan, Magnetic nanocellulose: A potential material for removal of dye from water, Journal of Hazardous Materials 394 (2020) 122571. https://doi.10.1016/j.jhazmat.2020.122571

[23] I.J.D. Ebenezar, S. Ramalingam, C.R. Raja, P.J. Prabakar, Vibrational spectroscopic [IR and raman] analysis and computational investigation [NMR, UV-Visible, MEP and kubo gap] on L-Valinium picrate, J. Nanotechnol. Adv. Mater 2 (2014) 11-25. https://doi.10.12785/jnam/020102

[24] V. Gopalakannan, N. Viswanathan, One pot synthesis of metal ion anchored alginate–gelatin binary biocomposite for efficient Cr (VI) removal, International Journal of Biological Macromolecules 83 (2016) 450-459. https://doi.10.1016/j.ijbiomac.2015.10.010

[25] A.K. Veeramachineni, T. Sathasivam, S. Muniyandy, P. Janarthanan, S.J. Langford, L.Y. Yan, Optimizing extraction of cellulose and synthesizing pharmaceutical grade carboxymethyl sago cellulose from Malaysian sago pulp, Applied Sciences 6 (2016) 170. https://doi.10.3390/app6060170

[26] N. Yacob, M.R. Yusof, Z.M.A. Ainun, K.H. Badri, Effect of cellulose fiber from sago waste on properties of starch-based films, In IOP Conference Series: Materials Science and Engineering 368 (2018) 012028. https://doi.10.1088/1757-899X/368/1/012028

[27] N.A. Rosli, I. Ahmad, I. Abdullah, Isolation and Characterization of Cellulose Nanocrystals from Agave angustifolia Fibre, Bioresources 8 (2013) 1893-1908

[28] H. Kargarzadeh, I. Ahmad, I. Abdullah, A. Dufresne, S. Y. Zainudin, R.M. Sheltami, Effects of hydrolysis conditions on the morphology, crystallinity, and thermal stability of cellulose nanocrystals extracted from kenaf bast fibers, Cellulose 19 (2012) 855-866. https://doi.10.1007/s10570-012-9684-6

[29] M.A. Al-Ghouti, D.A. Da'ana, Guidelines for the use and interpretation of adsorption isotherm models: A review, Journal of Hazardous Materials 393 (2020) 122383. https://doi.10.1016/j.jhazmat.2020.122383\

[30] S. Das, V.V. Goud, Characterization of a low-cost adsorbent derived from agro-waste for ranitidine removal, Materials Science for Energy Technologies 3 (2020) 879-888. https://doi.10.1016/j.mset.2020.10.009

[31] A. Rahman, T. Urabe, N. Kishimoto, Color removal of reactive procion dyes by clay adsorbents, Procedia Environmental Sciences 17 (2013) 270-278. https://doi.10.1016/j.proenv.2013.02.038

[32] X. Sun, X. Lv, C. Han., L. Bai, T. Wang, Y, Sun, Fabrication of Polyethyleneimine-Modified Nanocellulose/Magnetic Bentonite Composite as a Functional Biosorbent for Efficient Removal of Cu(II), Water 4 (2022) 2656. https://doi.10.3390/w14172656

[33] M. Kaur, S. Kumari, P. Sharma, Removal of Pb (II) from aqueous solution using nanoadsorbent of Oryza sativa husk: Isotherm, kinetic and thermodynamic studies, Biotechnology Reports 25 (2020) e00410. https://doi. 10.1016/j.btre. 2019.e00410

[34] Z. Ji, Y. Zhang, H. Wang, C. Li, Research progress in the removal of heavy metals by modified chitosan, Tenside Surfactants Detergents (2022) 2414. https://doi.10.1515/tsd-2021-2414

[35] G. Blanchard, M. Maunaye, G. Martin, Removal of heavy metals from waters by means of natural zeolites, Water research 18 (1984) 1501-1507. https://doi.10.1016/0043-1354(84)90124-6

[36] S. Azizian, Kinetic models of sorption: a theoretical analysis. Journal of colloid and Interface Science 276 (2004) 47-52. https://doi.10.1016/j.jcis.2004.03.048

Composite Materials: SEAJCCM2024
Materials Research Proceedings 56 (2025) 79-85

Materials Research Forum LLC
https://doi.org/10.21741/9781644903636-8

Effect of Microcrystalline Cellulose as a Filler on the Properties of Linear Low-Density Polyethylene/Thermoplastic Starch Blends

Nurnadia Farhana Che RAZALI[1] and Mohamad Kahar Ab WAHAB[1,a] *

[1]Faculty of Chemical Engineering & Technology, Universiti Malaysia Perlis, 06200 Arau, Perlis, Malaysia

[a]kaharwahab@unimap.edu.my

Keywords: Thermoplastic Starch, Microcrystalline Cellulose, Crystallite Size

Abstract. This study investigates the influence of microcrystalline cellulose (MCC) as a filler on the mechanical and structural properties of linear low-density polyethylene (LLDPE)/thermoplastic starch (TPS) blends. MCC was incorporated at varying concentrations (3%-12%) into LLDPE/TPS (70/30) using an internal mixer. The resulting composites were evaluated for their tensile properties, fracture morphology, functional group interactions, and crystallinity using a universal tensile testing machine, scanning electron microscopy (SEM), Fourier transform infrared spectroscopy (FTIR), and X-ray diffraction (XRD), respectively. The findings indicate that MCC incorporation enhances mechanical performance at lower concentrations, but excessive loading results in reduced tensile strength due to filler agglomeration. The study provides insights into optimizing MCC content for biodegradable polymer composites.

Introduction

The increasing environmental concerns regarding petroleum-based plastic waste have intensified research into biodegradable polymers. Conventional plastics, particularly polyethylene (PE), are widely used due to their durability and cost-effectiveness; however, their long degradation periods pose ecological challenges. Biodegradable alternatives, such as starch-based polymers, have gained attention due to their renewability and decomposition capability [1].

Polyethylene (PE), a semicrystalline thermoplastic with a high molecular weight and significant hydrophobicity, is one of the most widely used synthetic polymers. PE is available in three types: low-density polyethylene (LDPE), linear low-density polyethylene (LLDPE), and high-density polyethylene (HDPE). PE became widely utilized in the 1950s and was the first plastic material used for food packaging [2]. The production of plastics has significantly increased over the past few decades due to the rising consumer demand for this type of plastic. It is estimated that more than 100 million tonnes of plastic are produced worldwide each year [3]. This poses a major environmental problem due to the challenges in disposing of this plastic waste. Researchers have responded by developing biodegradable polymers made from natural resources. Biodegradable polymers can be broken down by microorganisms such as bacteria, algae, fungi, and others, meaning they do not harm the environment when used. To produce biodegradable plastic, synthetic and natural materials are blended together. The use of natural resources for biodegradable plastic offers several advantages: they are abundant, reasonably priced, readily available, and can decompose in the environment [4]. Common natural materials used as bio-based polymers include cellulose, chitosan, and starch.

Starch is a widely used polysaccharide globally. By mixing starch with a plasticizer at high temperature and shear, it can be converted into a true plastic known as thermoplastic starch (TPS) [5]. Blending TPS with other synthetic polymers enhances biodegradability while also providing high ductility, high modulus with toughness, and low cost [6]. However, a drawback of blending with TPS is the reduction in the mechanical properties of the blends due to the compatibility issues

Content from this work may be used under the terms of the Creative Commons Attribution 3.0 license. Any further distribution of this work must maintain attribution to the author(s) and the title of the work, journal citation and DOI. Published under license by Materials Research Forum LLC.

Composite Materials: SEAJCCM2024　　　　　　　　　　　　　　　Materials Research Forum LLC
Materials Research Proceedings 56 (2025) 79-85　　　　　　　　https://doi.org/10.21741/9781644903636-8

between the synthetic polymer and starch [7]. Synthetic polymers are hydrophobic, whereas starch is hydrophilic, leading to poor interfacial adhesion between the two components.

To improve the mechanical properties, reinforcing fillers are added to the blends. In this study, microcrystalline cellulose (MCC) was used as a reinforcing filler. MCC is a naturally occurring compound derived from purified and partially depolymerized cellulose. It is a white powder with an average particle size of 20–80 μm, a low degree of polymerization, a high degree of crystallinity, and a large specific surface area, making it suitable for creating regenerated cellulose membranes [8]. This study primarily focuses on the effect of MCC on the properties of LLDPE/TPS blends.

Methodology

Materials
LLDPE was supplied by Lotte Chemical Titan, Malaysia, with a density of 0.92 g/cm³. To prepare thermoplastic starch (TPS), potato starch was blended with glycerol, which acted as a plasticizer. The food-grade potato starch was supplied by Thye Huat Chan Sdn. Bhd., glycerol was supplied by TTD Chemical, and microcrystalline cellulose (MCC) was supplied by Sigma-Aldrich.

Method
The composite samples were fabricated using a hot compression molding technique. The raw materials were precisely weighed and mixed using a mechanical stirrer at 140°C for 10 minutes to ensure homogeneous dispersion. The preparation of TPS was conducted using a heated two-roll mill, where 35 wt.% glycerol was thoroughly incorporated into 65 wt.% potato starch to achieve uniform plasticization. The resultant TPS was then blended with LLDPE at 140°C for an additional 10 minutes using an internal mixer, ensuring homogeneous dispersion of the polymer matrix. In the final stage, MCC was introduced in varying concentrations (3%-12%) and continuously mixed until a well-dispersed, homogenous composite was obtained, optimizing filler integration within the polymer matrix. The mixture was then transferred into a 15cm x 15cm rectangular mold. The molding process was conducted at 140°C under a pressure of 70 MPa for 10 minutes. After molding, the samples were allowed to cool at room temperature under controlled conditions to minimize internal stresses. To facilitate the removal of samples from the mold, silicon-based spray was applied before introducing the composite mixture. A summary of the composition of each fabricated sample is presented in Table 1.

Table 1. Composition of LLDPE/MCC composites.

Sample	LLDPE (%)	TPS (%)	MCC (%)
LLDPE/TPS 3%MCC	70	30	3
LLDPE/TPS 6%MCC	70	30	6
LLDPE/TPS 9%MCC	70	30	9
LLDPE/TPS 12%MCC	70	30	12

Characterizations
Tensile tests of the LLDPE/MCC composites were performed following ASTM D638, which specifies the specimen dimensions and testing conditions. The tensile tests were conducted using a universal testing machine with a crosshead speed of 10 mm/min, at ambient temperature. Scanning Electron Microscopy (SEM) with a JEOL JSM-6460LA model was employed to analyze the fracture surfaces and surface morphology of chosen samples. Fourier Transform Infrared Spectroscopy (FTIR) was used to analyze the structural composition of the samples, with measurements taken at wavelengths between 650 cm⁻¹ and 4000 cm⁻¹. The crystal structure of LLDPE/TPS, MCC powder, and LLDPE/TPS/MCC composites was analyzed using X-Ray Diffraction (XRD). Samples were scanned from 5° to 40° (2θ) at a scan rate of 1° min⁻¹. The

Composite Materials: SEAJCCM2024   Materials Research Forum LLC
Materials Research Proceedings 56 (2025) 79-85   https://doi.org/10.21741/9781644903636-8

resulting diffraction patterns were analyzed using X'Pert HighScore Plus software to determine crystal size and degree of crystallinity. The degree of crystallinity was calculated using Eq. 1 [10].

$$X_c = I_c/(I_c + I_a)$$ (1)

Where I_c refers to intensities of crystalline phase and I_a corresponding to intensities of amorphous phase. Meanwhile, the crystallite size of each sample was calculated using Scherrer method as given in Eq. 2 [11].

$$Crystallite\ size, L_c(nm) = K\lambda/\beta cos\theta$$ (2)

where
K = Dimensionless shape factor for spherical crystal particles (0.92)
λ = Wavelength of X-Ray (CuKα : 0.15406 nm)
β = Half high width of the diffraction peak (FWHM), rad
θ = Diffraction angle, rad

Results and Discussion

Tensile Properties
From the results obtained, the LLDPE/TPS 70/30 blend exhibited a tensile strength of 10.57 MPa and an elongation at break of 265.3%. Based on Fig. 1, the tensile strength of LLDPE/TPS blends with 3% MCC increased slightly to 10.889 MPa. However, increasing the MCC content from 6% to 12% resulted in a decrease in both tensile strength and elongation at break. The initial enhancement in composite strength is attributed to effective filler–matrix interactions, which facilitate greater stress transfer from the matrix to the filler during external loading. According to previous studies, increasing filler loading leads to the formation of filler aggregates composed of particles held together by weak forces. These aggregates are easily debonded from the matrix, resulting in a decrease in the composite's tensile strength [6].

The increase in strength of the LLDPE/TPS composite at lower MCC loadings is attributed to the homogeneous dispersion of MCC particles. Well-dispersed particles extend the crack propagation path, absorb a portion of the energy, and enhance plastic deformation. However, with increased MCC loading, the detachment of filler particles from the LLDPE/TPS matrix creates voids that facilitate crack formation. Additionally, the agglomeration of MCC reduces mechanical strength due to the inherently low strength of the agglomerates themselves.

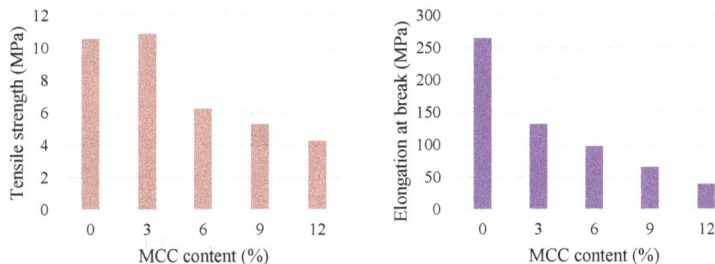

Fig. 1. Tensile strength and elongation at break of LLDPE/TPS/MCC composites.

Composite Materials: SEAJCCM2024
Materials Research Proceedings 56 (2025) 79-85

Materials Research Forum LLC
https://doi.org/10.21741/9781644903636-8

Morphological Analysis

Fig. 2(a–d) show the SEM images of LLDPE/TPS, MCC powder, and LLDPE/TPS/MCC composites. The image at 500x magnification reveals significant voids at the interphase between LLDPE and TPS. These voids in the LLDPE/TPS blends indicate a poor degree of interfacial interaction between LLDPE and TPS, primarily due to the compatibility issues arising from the hydrophilic nature of TPS and the hydrophobic nature of LLDPE [4]. As a result, TPS particles do not disperse homogeneously within the LLDPE matrix [6].

Conversely, as shown in Figures 2c and 2d, the SEM micrographs of the LLDPE/TPS/MCC composites with 3% MCC exhibit a more robust interfacial phase compared to the LLDPE/TPS blends. However, in the LLDPE/TPS/MCC composites with 9% MCC, agglomeration is observed. This may be due to the higher MCC content, which does not disperse well within the LLDPE/TPS matrix, leading to agglomeration and poor blend uniformity.

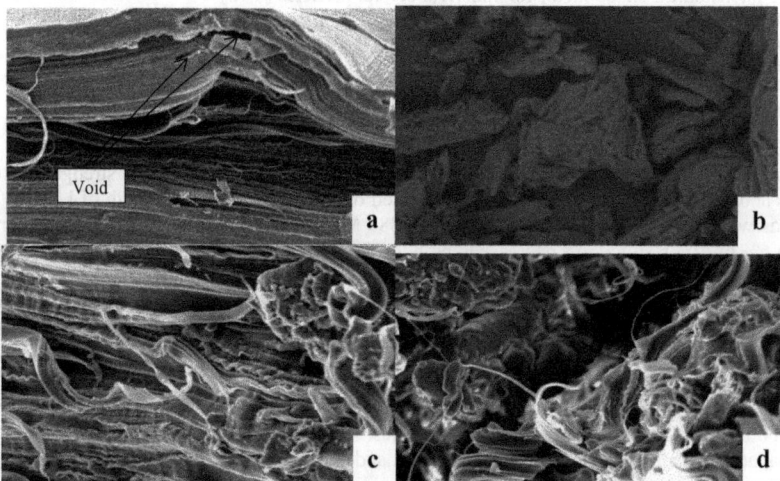

Fig. 2. SEM micrograph for (a) LLDPE/TPS, (b) MCC powder, (c) LLDPE/TPS/MCC 3% and (d) LLDPE/TPS/MCC 9%.

Structural Analysis (Fourier-Transform Infrared Spectroscopy)

Fig. 3 presents the observation of functional groups in biodegradable plastics using infrared spectroscopy. The functional group analysis was conducted to identify the functional groups present in the compounds. The results indicated the presence of characteristic absorption peaks for LLDPE, specifically in the range of 2909–2846 cm^{-1}, corresponding to the CH stretching vibrations of LLDPE [4]. The peak at 1022 cm^{-1} was attributed to C–O–C bond bending, while the peak at 3270 cm^{-1} was associated with O–H bond stretching, characterizing the functional groups in TPS.

Additionally, the characterization peaks for the O–H bond and C–O–C bond exhibited distinct spectral peaks, suggesting possible interactions between starch and glycerol [9]. The peak at 1458 cm^{-1} was attributed to C–H$_2$ bending, which is related to the crystalline or amorphous regions in certain cellulosic molecules [10].

Composite Materials: SEAJCCM2024   Materials Research Forum LLC
Materials Research Proceedings 56 (2025) 79-85   https://doi.org/10.21741/9781644903636-8

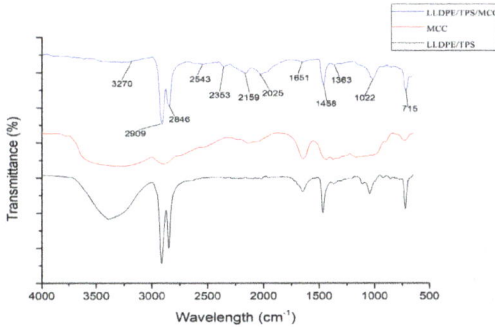

Fig. 3. FTIR analysis for MCC powder, LLDPE/TPS blends and LLDPE/TPS/MCC composite.

X-Ray Diffraction Analysis

Fig. 4 presents the X-ray diffraction patterns of MCC powder, LLDPE/TPS blends, and LLDPE/TPS/MCC composites. The characteristic peaks were scanned over a 2θ range from 5° to 40°. The results show that the intensity peaks for the LLDPE/TPS blends appeared at 19.2°, 21.3°, 22.0°, 23.6°, 29.7°, 33.7°, 36.0°, and 39.4°. The MCC powder exhibited peaks at 9.2°, 14.6°, 22.5°, and 34.4°. For the LLDPE/TPS/MCC composites, peaks were observed at 10.21°, 20.2°, 22.2°, 24.5°, 30.7°, 34.6°, and 36.9°. Based on these XRD results, the crystallite size and degree of crystallinity of the samples were calculated.

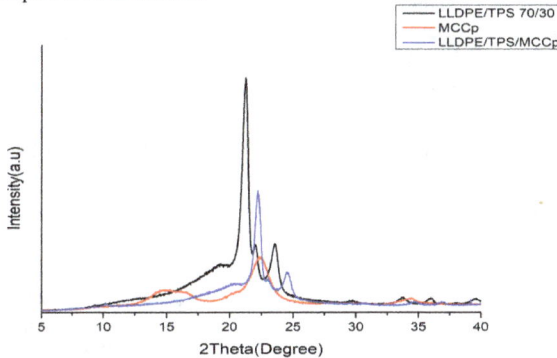

Fig. 4. XRD analysis of MCC powder, LLDPE/TPS blends and LLDPE/TPS/MCC polymer composite.

According to the Table 2, MCC powder (MCCp) exhibits the largest crystallite size, followed by the LLDPE/TPS blend and the LLDPE/TPS/MCC polymer composites. As crystallite size decreases, the diffraction peaks become broader [13]. The addition of MCC powder to the blends results in an increase in the degree of crystallinity of the polymer blend. Higher crystallinity enhances the strength of the polymer, as the intermolecular bonding in the crystalline phase is more pronounced. Consequently, this increased crystallinity leads to higher polymer strength, as the deformation of the polymer results in more oriented chains [14].

Table 2. Crystallite size and degree of crystallinity for LLDPE/TPS, MCCp and LLDPE/TPS/MCC polymer composite.

Sample	Crystallite size (nm)	Degree of crystallinity (%)
LLDPE/TPS	14.800	51.051
MCCp	29.325	59.689
LLDPE/TPS/MCC	14.286	53.823

Conclusions

Based on the results of this study, it can be concluded that MCC significantly influences the mechanical properties of LLDPE/TPS blends. Various characterizations, including tensile properties, SEM micrographs, FTIR analysis, and XRD analysis, indicate that LLDPE/TPS/MCC polymer composites show promise for future bio-based plastic packaging applications. The incorporation of 3% MCC into the LLDPE/TPS blend enhances the strength of the composites. Additionally, the presence of starch and cellulose—both natural materials—increases the biodegradability of the composites. Bio-based polymers offer considerable advantages for future applications, including environmental benefits and improved cost-effectiveness and processability.

Acknowledgements

The author would like to acknowledge the support from the Fundamental Research Grant Scheme (FRGS) under a grant number of (Ref: FRGS/1/2019/TK05/UNIMAP/02/13) from the Ministry of Higher Education Malaysia.

References

[1] Y.A. Mubarak, R.T. Abdulsamad, Effects of microcrystalline cellulose on the mechanical properties of low-density polyethylene composite, J. Thermo. Compost. Mater. 32 (2018) 297-311. https://doi.org/10.1177/0892705717753056

[2] W. Waryat, M. Romli, A. Suryani, I. Yuliasih, S. Johan, Using of a Compatibilizer to Improve Morphological, Physical and Mechanical Properties of Biodegradable Plastic From Thermoplastic Starch/LLDPE Blends, Inter. J. Eng. Technol. 14 (2018) 131218. https://doi.org/10.17146/jsmi.2013.14.3.742

[3] A.A.H. Anis, A.W.M. Kahar, Preparation and characterization of LDPE/TPS blend filled with calcium carbonate, IOP Conf. Ser.: Mater. Sci. Eng. 957 (2020) 012015. https://doi.org/10.1088/1757-899X/957/1/012015

[4] A.W.M. Kahar, L.J. Ann, Preparation and Characterisation of Linear Low-Density Polyethylene / Thermoplastic Starch Blends Filled with Banana Fibre, IOP Conf. Ser.: Mater. Sci. Eng. 209 (2017) 012003. https://doi.org/10.1088/1757-899X/209/1/012003

[5] M. Li, T. Wei, C. Qian, Z. Liang, Preparation of microcrystalline cellulose from Rabdosia rubescens residue and study on its membrane properties, Sci. Rep. 11 (2021) 18956. https://doi.org/10.1038/s41598-021-98645-x

[6] A. Martins, A.K. Cattelan, R. Santana, How the compatibility between polyethylene and thermoplastic starch can be improved by adding organic acids?, Polym. Bull. 75 (2018) 2197-2212. https://doi.org/10.1007/s00289-017-2147-3

[7] K. Juntira, C.T.N. Luangsa-ard, Characteristics of starch-filled LLDPE plastic processed from plastic waste, J. Chin. Inst. Eng. 35 (2012) 45-50. https://doi.org/10.1080/02533839.2012.624817

[8] N. Amigo, H. Palza, D. Canales, F. Sepúlveda, D.A. Vasco, F. Sepúlveda, P.A. Zapata, Effect of starch nanoparticles on the crystallization kinetics and photodegradation of high-

density polyethylene, Compos. Part B. 174 (2019) 106979.
https://doi.org/10.1016/j.compositesb.2019.106979

[9] A.H.M. Zain, M.K.A. Wahab, H. Ismail, Effect of calcium carbonate incorporation on the properties of low linear density polyethylene/thermoplastic starch blends, J. Eng. Sci. 15 (2019) 97-108.

[10] N. Atykyan, V. Revin, V. Shutova, Raman and FT-IR Spectroscopy investigation the cellulose structural differences from bacteria Gluconacetobacter sucrofermentans during the different regimes of cultivation on a molasses media, AMB Exp. 10 (2020) 84.
https://doi.org/10.1186/s13568-020-01020-8

[11] S. Fatimah, R. Ragadhita, D.F. Al Husaeni, A.B.D. Nandiyanto, How to calculate crystallite size from x-ray diffraction (XRD) using Scherrer method, ASEAN J. Sci. Eng. 2 (2022) 65-76.
https://doi.org/https://ejournal.kjpupi.id/index.php/ajse/article/view/283

[12] M. Doumeng, L. Makhlouf, F. Berthet, O. Marsan, K. Delbé, L. Denape, F. Chabert, A comparative study of the crystallinity of polyetheretherketone by using density, DSC, XRD, and Raman spectroscopy techniques, Polym. Test. 93 (2021) 106878.
https://doi.org/10.1016/j.polymertesting.2020.106878

[13] S.A. Speakman, Estimating crystallite size using XRD, MIT Center Mater. Sci. Eng. 2 (2014) 3-8.

[14] A. Galeski, Strength and toughness of crystalline polymer systems, Progr. Polym. Sci. 28 (2003) 1643-1699. https://doi.org/10.1016/j.progpolymsci.2003.09.003

Composite Materials: SEAJCCM2024
Materials Research Proceedings 56 (2025) 86-92

Materials Research Forum LLC
https://doi.org/10.21741/9781644903636-9

The Study on Different Percentages of Nickel (Ni) Addition on Sn0.7Cu Solder Alloy Properties

Athirah Khemar[1,a], Kamrosni Abdul Razak[1,b]*, Nur Izzati Muhammad Nadzri[1],
Mohd Arif Anuar Mohd Salleh[1], Faiz Azri Fadlollah[1]

[1]Centre of Excellence Geopolymer and Green Technology, Faculty of Chemical Engineering Technology, Universiti Malaysia Perlis (UniMAP), Taman Muhibah, 02600, Jejawi, Arau, Perlis, Malaysia

[a]athirahkhemar@studentmail.unimap.edu.my, [b]kamrosni@unimap.edu.my

Keywords: Sn0.7Cu, Microstructure, Intermetallic Compound (IMC) Layer, Solder Joint

Abstract. The rapid development of electronic products has created challenges for designers and manufacturers in selecting the best electronic packaging materials for their products. Among these materials, lead-free solders have emerged as the best replacement for lead-containing solders, despite the fact that their performance still falls short of the original specification. The main of this research is to investigate the effect of Nickel (Ni) addition with different percentages into Sn0.7Cu solder alloy on the intermetallic compound formation (IMC) and growth at the solder joint interface. Based on weight percentage (wt.%) changes, the work was carried out by producing Sn0.7Cu-xNi (x = 0.0, 0.03, 0.05, 0.07). The type of IMC produced after reflowing when increasing the nickel addition in SnCu solder was decreased in thickness but smoother and finer. This study reveals that Ni as an excellent candidate for improving solder joint performance for Cu_6Sn_5 layer to form into $(Cu,Ni)_6Sn_5$ layer. Furthermore, the solder joint strength was found to be increased by increasing the Ni percentage which make Ni as a good candidate for improving the solder joint which focused for IMC layer formation.

Introduction

Soldering is a procedure in which two or more metal items are connected by melting and inserting a solder filler metal into the joint, the solder having a lower melting point than the neighboring metal [1]. During soldering the solder alloy transitions into a melt state which interacts with the substrates to generate intermetallic compounds that form between the solder and the substrate material. The solder alloy reacts as a fusible metallic mixture that enables permanent component connections between different metal parts. To bind and link the components together, the solder must be heated. In addition, the solder should be resistant to oxidative and corrosive actions that would weaken the junction over time [2].

An intermetallic compound is any of a group of compounds that include two or more elemental metals in fixed ratios as opposed to continually variable combinations (as in solid solutions)[3]. Intermetallic combinations' crystal structures and characteristics can differ significantly from those of their components. The relative sizes of the atoms and the ratio of the total number of valence electrons to the total number of atoms, in addition to the typical valences of their components, have a significant impact on the composition of intermetallic compounds [4]. Due to environmental concerns, the European Union (EU) has been enacting and enforcing regulations to limit the use of harmful elements, such as lead (Pb), in electronic devices since 2002 [5]. Lead toxicity has prompted the development of a suitable lead-free solder alloy with low cost, good wettability, and acceptable physical, mechanical, metallurgical, and fatigue-resistant qualities Tin-silver-copper (SAC), tin-3 copper (SnCu), tin-silver (SnAg), and tin zinc (SnZn) are the most often used lead-free solder alloys in the microelectronic sector [6].

Content from this work may be used under the terms of the Creative Commons Attribution 3.0 license. Any further distribution of this work must maintain attribution to the author(s) and the title of the work, journal citation and DOI. Published under license by Materials Research Forum LLC.

Composite Materials: SEAJCCM2024   Materials Research Forum LLC
Materials Research Proceedings 56 (2025) 86-92   https://doi.org/10.21741/9781644903636-9

Sn0.7Cu solders are among the emerging Pb-free alloy possibilities that are both affordable and promising. Sn0.7Cu solder has been frequently employed in wave soldering processes as the practicability of Sn-0.7Cu solder has been consistently demonstrated [7]. The Sn-Cu-Ni alloy, which has been used in wave, dip, and iron soldering methods, is one appealing lead-free solder since it eliminates the usage of pricey silver or rare earth elements found in competing alloys. Previous studies have demonstrated that Sn-0.7Cu-0.05Ni has better fluidity and oxidation resistance than Sn0.7Cu [8].

The Sn0.7Cu solder alloy is widely used in electronic applications; however, its microstructure and mechanical properties can be further optimized. The addition of Ni has been proposed to refine the eutectic region and β-Sn morphology, potentially enhancing the alloy's performance. However, excessive Ni content may lead to the formation of intermetallic compound (IMC) dendrites, which can increase brittleness over time. Therefore, it is crucial to investigate the effects of varying Ni percentages on the microstructural evolution and mechanical reliability of Sn0.7Cu solder joints. This study focused on the addition of different percentages of Ni into the Sn0.7Cu solder alloy. The analysis of raw material used the X-ray diffraction (XRD) method. The microstructure and intermetallic compound were observed under an optical microscope to study the morphology of the samples.

Methodology

Raw materials preparation

The raw materials for the Sn0.7Cu solder alloy were supplied by Nihon Superior (M) Sdn. Bhd, Ipoh, Perak, Malaysia. The solder alloy was prepared by melting the ingots in a solder pot at 250°C. The substrate used was a copper sheet with 99.9% purity, measuring 15 mm × 15 mm with a thickness of 0.3 mm. Nickel addition was carried out using the casting method, where Sn0.7Cu solder alloy was melted at 550°C, and nickel particles were introduced into the molten alloy. Three different weight percentages of nickel 0.03 wt.%, 0.05 wt.%, and 0.07 wt.% were added. The molten mixture was then poured onto a ceramic plate to cool, forming Sn0.7Cu0.03Ni, Sn0.7Cu0.05Ni, and Sn0.7Cu0.07Ni, which were used as solder materials.

Sample preparation

A thin sheet soldering method was used to prepare Sn0.7Cu, Sn0.7Cu0.03Ni, Sn0.7Cu0.05Ni, and Sn0.7Cu0.07Ni solder joints. Alloys were rolled into thin sheets with an approximate thickness of ~35 μm and then cut into 10 mm × 10 mm pieces to be placed onto the cleaned substrate. For substrate preparation, a copper (Cu) sheet was used and cleaned thoroughly. The cleaning process involved immersing the substrate in 5% hydrochloric acid (HCl) to remove the oxidized layer, followed by washing with distilled water at room temperature. The substrate was then cleaned again using acetone (C3H6O), rinsed with distilled water, and air-dried. After drying, flux paste was applied to the substrate to facilitate soldering. Next, the thin sheets of Sn0.7Cu, Sn0.7Cu0.03Ni, and Sn0.7Cu0.05Ni were placed onto the flux-coated substrate, with an additional layer of flux paste applied on top of the solder alloy to enhance joint formation during reflow. The samples were then subjected to reflow soldering in a reflow oven at 250°C for 300 seconds. After reflow, the samples were cleaned in an ultrasonic bath with acetone for 5 minutes and left at room temperature for 24 hours to improve the microstructure of the solder joint and reduce residual stress in the IMC layer. Finally, the samples were mounted and polished for an initial study of the IMC layer with and without nickel addition.

Bulk solder and intermetallic compound (IMC) characterization

For bulk solder, X-ray diffraction (XRD) analysis was conducted to observe phase formation, while microstructure analysis was performed using an optical microscope. For IMC characterization, reflowed samples were ground and polished for morphology observation. The

Composite Materials: SEAJCCM2024   Materials Research Forum LLC
Materials Research Proceedings 56 (2025) 86-92   https://doi.org/10.21741/9781644903636-9

cross-sections of the samples were also examined under an optical microscope, and the average IMC thickness was measured using Eq. 1.

$$\text{Thickness}, t = \frac{\text{Area}, A}{\text{Length}, L} \tag{1}$$

Results and Discussion

Effect of Ni addition to the phase formation

The X-ray diffraction (XRD) analysis in Fig. 1 demonstrates the main peaks of Cu_6Sn_5 within the solder alloy. Raising Ni concentration in Sn0.7Cu solder induced peak height variations in the Cu_6Sn_5 phase based on the specific Ni content level. The addition of Ni creates an ideal condition for turning Cu_6Sn_5 into $(Cu,Ni)_6Sn_5$. Refractometric analysis confirmed the formation of $(Cu,Ni)_6Sn_5$ when subjecting the specimens to Ni additions at 0.05 wt.% and 0.07 wt.% yet failed to manifest at 0.03 wt.% Ni. Through x-ray diffraction scanning Cu_6Sn_5 formation displayed its peaks between 26.86° to 90° with maximum intensity at 30° as the crystal structure matched monoclinic Cu_6Sn_5. The X-ray diffraction pattern revealed a peak at 45° for both Sn0.7Cu0.05Ni and Sn0.7Cu0.07Ni samples that was identified as $(Cu,Ni)_6Sn_5$. The existing literature demonstrates that Cu_6Sn_5 shows monoclinic or hexagonal phase behavior following thermal expansion variations [4, 9].

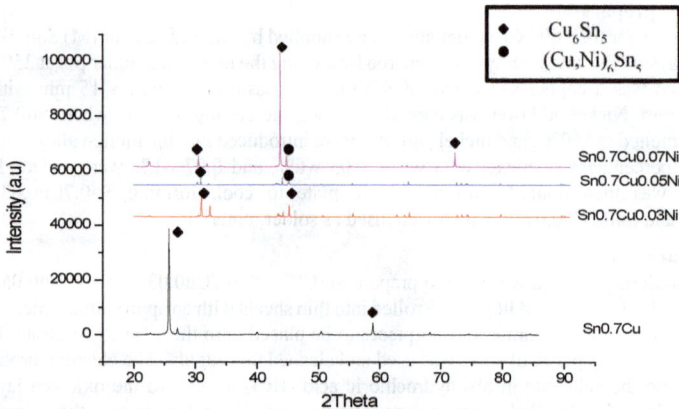

Fig. 1. X-Ray diffraction peak profile comparing the different Ni percentages addition in Sn0.Cu solder alloy.

Effect of Ni Addition on Microstructure and IMC Morphology

The addition of Ni into Sn0.7Cu solder found changes in the surface formation as referred to in Fig. 2 focused on the eutectic region and the beta-tin morphology on the bulk solder. Because Ni shared the same atomic radius and similar properties as a metal element adjacent to Cu, it was suitably used as the additional element for modifying the Sn0.7Cu alloy. Since Ni has the same atomic radius and similar properties as a metal element adjacent to Cu therefore, Ni is suitably used as the additional element for the modification of Sn0.7Cu alloy. In general, Ni particles can replace the Cu atom in the solder alloy and a small amount of Ni can form a new phase from Cu_6Sn_5 to $(Cu,Ni)_6Sn_5$ as the primary phase and refine the primary β-Sn phase. From previous studies, a cellular-type eutectic microstructure in Sn-0.7Cu0.05Ni by solder casting technique [4], the results showed increasing Ni percentages content, and the average grain size was decreasing especially for the β-Sn phase where more eutectic regions were observed on the microstructure of Sn0.7Cu Fig. 2(a) and Sn0.7Cu0.03Ni Fig. 2(b). Meanwhile, the phase of $(Cu,Ni)_6Sn_5$ formed as the increasing of the Ni percentages where it improved the Sn0.7Cu alloy as in Fig. 2(c). The effect of the high contain of Ni in the Sn0.7Cu solder caused the formation of dendrites on the surface morphology where the dendrites in the IMC components of the solder alloy in Fig. 2(d) which contain 0.07wt% of Ni. The IMC can cause brittleness to the solder if in prolonged use [10].

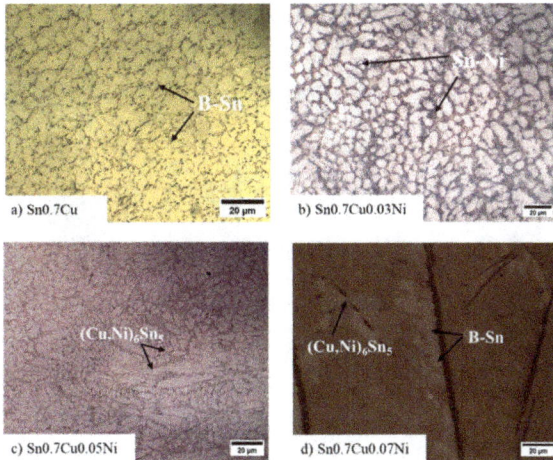

Fig. 2. Microstructure of bulk sample of the solder.

The nickel addition into Sn0.7Cu solder alloy has made changes to the morphology of the IMC layer of the material. Sn0.7Cu-xNi solder alloy are different weight percentages added between 0.03 wt.%, 0.05 wt.% and 0.07 wt.%. The cross-sectional view of solder Sn0.7Cu, Sn0.7Cu0.03Ni, Sn0.7Cu0.05Ni and Sn0.7Cu0.07Ni lead to the study on the IMC layers formation. The changes that occur in the microstructure of the solder depend on the material composition added as to study the mechanical properties and the surface morphology of the samples. As in Fig. 3, different of Ni contained have resulted in changes in its IMC layer thickness morphology after reflow. It was observed that the sample caused the formation of the Cu_6Sn_5 layer which was the fastest IMC form since it provides a fast diffusion passageway for Cu atoms diffused towards the main solder alloy which also has been discovered from past studies [11, 12]. The IMC layer thickness was found to be increased after the increment of Ni percentage in the Sn0.7Cu solder alloy where results showed the IMC thickness decreasing after reflowing at 250°C as stated in Fig. 4.

The cross-sectional of the IMC layer can be seen in Fig. 3, where the changes of the IMC layer for Sn0.7Cu show a scallop shape structure for Figure 3(a and b), meanwhile for Fig. 3(c and d) resulting in a flat shape IMC layer. The thickness from Sn0.7Cu, Sn0.7Cu0.03Ni, Sn0.7Cu0.05Ni and Sn0.7Cu0.07Ni shows a decreasing trend for the thickness as the influence of the different percentages of Ni has changed the properties of the IMC layer to occur in short time. Moreover, the addition of Ni into Sn0.7Cu solder alloy converted the features of the IMC from scallop morphology to flat morphology. Thus, for further studies longer time is needed to enhance the IMC layer thickness [13]. This also shows that the addition of Ni into Sn0.7Cu solder alloys leads to form a better material for industrial applications nowadays. Fig. 4 shows the average thickness with in decreasing trend as during the reflow process the dissolution thickness of the Cu substrate

Composite Materials: SEAJCCM2024
Materials Research Proceedings 56 (2025) 86-92

Materials Research Forum LLC
https://doi.org/10.21741/9781644903636-9

for Sn0.7Cu-xNi solder was the minimum for all samples. The addition of Ni to Sn0.7Cu solder can be effective in reducing the formation of IMC layers on the interface during the reflow process and also can exhibit IMC growth which aligned with past studies [14]. The thickness from Sn0.7Cu, Sn0.7Cu0.03Ni, and Sn0.7Cu0.07Ni shows a decreasing trend for the thickness since the influence of the different percentages of Ni has changed the properties of the IMC layer to occur in a short time as longer time needed to enhance the IMC layer thickness. However, the differences showed for sample Sn0.7Cu0.05Ni where the thickness is increased with a thickness average of 0.9284 μm because there is the probability of the Ni was not fully solidified during the casting process also minimal Ni contained in the solder alloy [15].

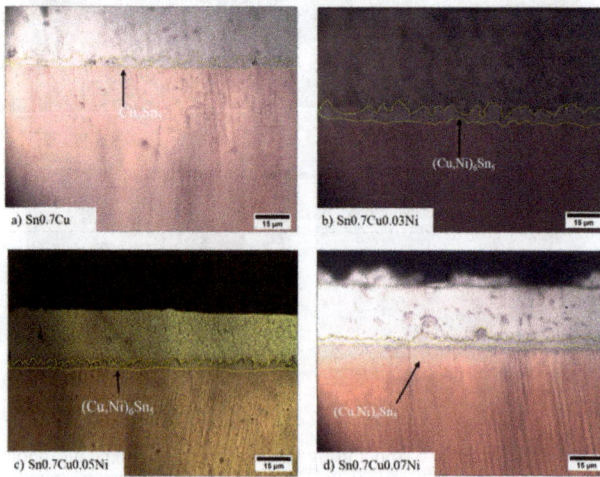

Fig. 3. IMC thickness at a magnification of 40X.

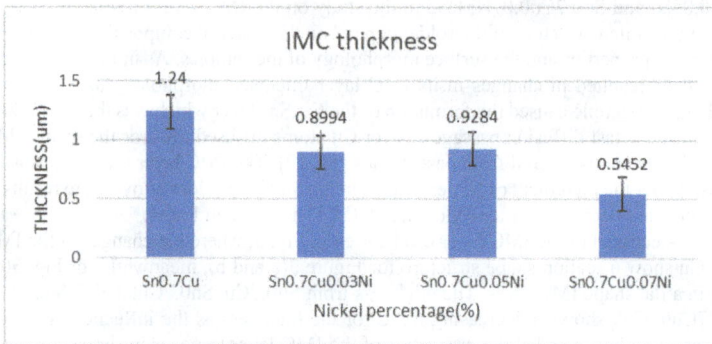

Fig. 4. Comparison of IMC thickness cross-section view.

Conclusion
This study demonstrates that adding Ni to Sn0.7Cu solder alloy influences the formation of intermetallic compounds (IMCs) and improves solder joint performance. The transformation of Cu_6Sn_5 into $(Cu,Ni)_6Sn_5$ was observed at Ni additions of 0.05 wt.% and 0.07 wt.%, refining the

microstructure and reducing IMC layer thickness. While Ni enhances solder joint strength and stability, excessive Ni content can lead to dendritic IMC structures, which may increase brittleness over time. Therefore, optimizing the Ni content in Sn0.7Cu solder is crucial for achieving better mechanical properties and long-term reliability in electronic applications.

Acknowledgements
The author would like to acknowledge the support from the Fundamental Research Grant Scheme (FRGS) under a grant number of FRGS/1/2020/TKO/UNIMAP/03/18 from the Ministry of Higher Education Malaysia also Faculty of Chemical Engineering Technology for the materials, facilities and finance support.

References
[1] Y. Li, C. Chen, R. Yi, Y. Ouyang, Special brazing and soldering, Journal of Manufacturing Processes 60 (2020) 608-635. https://doi.org/10.1016/J.JMAPRO.2020.10.049

[2] Information on https://www.mechdaily.com/what-is-annealing/

[3] P.T. Vianco, A review of interface microstructures in electronic packaging applications: Soldering technology, Jom 71 (2019) 158-177. https://doi.org/10.1007/s11837-018-3219-z

[4] M. Zhao, L. Zhang, Z.Q. Liu, M.Y. Xiong, L. Sun, Structure and properties of Sn-Cu lead-free solders in electronics packaging, Science and technology of advanced materials 20 (2019) 421-444. https://doi.org/10.1080/14686996.2019.1591168

[5] N.A. Ezaham, N.R.A. Razak, M.A.A.M. Salleh, Influence of bismuth on the solidification of tin copper lead-free solder alloy, AIP Conference Proceedings 2045 (2018) 020104. https://doi.org/10.1063/1.5080917

[6] G. Zeng, S.D. McDonald, D. Mu, Y. Terada, H. Yasuda, Q. Gu, M.A.A. Mohd Saleh, K. Nogita, The influence of ageing on the stabilisation of interfacial (Cu, Ni) 6 (Sn, Zn) 5 and (Cu, Au, Ni) 6Sn5 intermetallics in Pb-free Ball Grid Array (BGA) solder joints, Journal of Alloys and Compounds 685 (2016) 471-482. https://doi.org/10.1016/j.jallcom.2016.05.263

[7] N. Adli, N.R. Abdul Razak, N. Saud, Physical and mechanical behaviors of SnCu-based lead-free solder alloys with an addition of aluminium, Applied Mechanics and Materials 815 (2015) 64-68. https://doi.org/10.4028/www.scientific.net/amm.815.64

[8] Information on https://blog.matric.com/lead-vs-lead-free-solder-in-pcb-manufacturing#:~:text=Lead%2Dfree%20solder%20is%20a,use%20of%20lead%2Dbased%20solder

[9] D. Mu, J. Read, Y. Yang, K. Nogita, Thermal expansion of Cu 6 Sn 5 and (Cu, Ni)6Sn5, Journal of Materials Research 26 (2011) 2660-2664. https://doi.org/10.1557/jmr.2011.293

[10] S.U. Mehreen, K. Nogita, S. McDonald, H. Yasuda, D. StJohn, Suppression of Cu3Sn in the Sn-10Cu peritectic alloy by the addition of Ni, Journal of Alloys and Compounds 766 (2018) 1003-1013. https://doi.org/10.1016/j.jallcom.2018.06.251

[11] J. Wang, C. Leinenbach, H.S. Liu, L.B. Liu, M. Roth, Z.P. Jin, Diffusion and atomic mobilities in fcc Ni-Sn alloys, Journal of phase equilibria and diffusion 31 (2010) 28-33. https://doi.org/10.1007/s11669-009-9607-x

[12] E. Efzan, A. Marini, A review of solder evolution in electronic application, International Journal of Engineering 1 (2012) 2305-8269

Composite Materials: SEAJCCM2024
Materials Research Proceedings 56 (2025) 86-92

Materials Research Forum LLC
https://doi.org/10.21741/9781644903636-9

[13] A.N. Hashim, M.A.A.M. Salleh, M.M. Ramli, M.M.A.B. Abdullah, A.V. Sandu, P. Vizureanu, I.G. Sandu, Effect of isothermal annealing on Sn whisker growth behavior of Sn0. 7Cu0. 05Ni solder joint, Materials 16 (2023) 1852. https://doi.org/10.3390/ma16051852

[14] H. Nishikawa, J.Y. Piao, T. Takemoto, Interfacial reaction between Sn-0.7 Cu (-Ni) solder and Cu substrate, Journal of Electronic Materials 35 (2006) 1127-1132. https://doi.org/10.1007/BF02692576

[15] L. Somlyai-Sipos, P. Baumli, Effect of nickel addition on the wettability and reactivity of tin on copper substrate, Resolution and Discovery 2 (2017) 9-12. https://doi.org/10.1556/2051.2017.00039

Composite Materials: SEAJCCM2024   Materials Research Forum LLC
Materials Research Proceedings 56 (2025) 93-100   https://doi.org/10.21741/9781644903636-10

Physical and Mechanical Properties of Steel-Polypropylene Composite Fiber Geopolymer Concrete

Meor Ahmad FARIS[1,2,a] *, Warid Wazien AHMAD ZAILANI[3,b],
Muhammad Fahem MOHD TAHIR[2,4,c], Mohammad Firdaus ABU HASHIM[1,2,d],
M.N.A. UDA[1,5,e], Mohd Ikram RAMLI[6,f], Mohammed Izzuddeen MOHD YAZID[7,g]

[1]Faculty of Mechanical Engineering Technology, Universiti Malaysia Perlis (UniMAP), Pauh Putra 026000 Arau, Perlis, Malaysia

[2]Center of Excellence Geopolymer and Green Technology (CEGeoGTech), Universiti Malaysia Perlis, 01000 Kangar, Perlis, Malaysia

[3]School of Civil Engineering, College of Engineering, Universiti Teknologi MARA (UiTM), 40450 Shah Alam, Malaysia

[4]Faculty of Chemical Engineering Technology, University Malaysia Perlis, 01000 Kangar, Perlis, Malaysia

[5]Institute of Nano Electronic Engineering, Universiti Malaysia Perlis, 01000 Kangar, Perlis, Malaysia

[6]School of Engineering, Universiti of Wollongong Malaysia, Jalan Kontraktor U1/14, Seksyen U1, 40150 Shah Alam, Malaysia

[7]Faculty of Innovative Design & Technology, Universiti Sultan Zainal Abidin, 21030 Kuala Terengganu, Malaysia

[a]meorfaris@unimap.edu.my, [b]waridwazien@uitm.edu.my, [c]faheem@unimap.edu.my, [d]firdaushashim@unimap.edu.my, [e]nuraiman@uitm.edu.my, [f]ikram.r@uow.edu.my, [g]mohdizzuddeen@unisza.edu.my

Keywords: Geopolymer, Steel Fiber, Polypropylene Fiber, Reinforced Concrete, Hybrid Fibers

Abstract. The manufacturing of cement for use in construction sites all over the world has resulted in tonnes of carbon dioxide being released. It is better to replace an alternative material such as geopolymer which contributes less carbon footprint than traditional Portland cement. Concrete is the most versatile building material, yet it has drawbacks in mechanical and physical properties, such as limited ductility, high water absorption, and low compressive strength. This study aims to determine the effect of the addition of composite fibers on the mechanical and physical properties of geopolymer concrete. In this research, the physical and mechanical properties of geopolymer concrete were investigated by mixing Class F fly ash with an alkaline activator consisting of sodium hydroxide and sodium silicate. Steel fiber and polypropylene cut into 2 mm fiber were added into the geopolymer concrete as reinforcement. Various volume percentages ranging from 0% to 2% are used. Density, water absorption, workability, and compression testing were performed on all geopolymer concrete reinforced with steel and polypropylene fiber with varying volume percentages. The density of geopolymer concrete is similar to that of Ordinary Portland Cement (OPC), which is around 2400 kg/m^3, and it has gradually increased with the inclusion of steel fiber, while more polypropylene fiber causes lesser density. With an increment of steel fiber and a decrement of polypropylene fiber, the result of water absorption percentage shows a decrement. Besides, the inclusion of steel fibers reduces the workability of geopolymer concrete.

Content from this work may be used under the terms of the Creative Commons Attribution 3.0 license. Any further distribution of this work must maintain attribution to the author(s) and the title of the work, journal citation and DOI. Published under license by Materials Research Forum LLC.

Composite Materials: SEAJCCM2024
Materials Research Proceedings 56 (2025) 93-100

Materials Research Forum LLC
https://doi.org/10.21741/9781644903636-10

However, the addition of polypropylene fibers shows a higher workability. Plus, the inclusion of steel fiber and the reduction of polypropylene fiber improve the compressive strength.

Introduction

Geopolymer is an inorganic polymer made out of tetrahedral three-dimensional aluminium silicate (AlO_4 and SiO_4). Geopolymer is made from amorphous cementitious material that has a lot of potential in terms of early compressive strength, volume stability, low permeability, high durability, acid resistance, and fire resistance [1-3]. Due to the use of recycled and underused materials such as fly ash, a byproduct of coal combustion, geopolymer production can contribute to lower costs and profitability. Geopolymer concrete is reported to be 10-30 % less expensive to produce than Ordinary Portland Cement (OPC) concrete [4]. Geopolymer cement has a lower cost because of the 50 % less energy used in the manufacturing process, which uses moderate temperatures (20 - 90 °C) compared to OPC cement, which uses much higher temperatures (1450 °C). Geopolymer based on fly ash has recently received a lot of attention [5-7].

Furthermore, geopolymer concrete, like OPC concrete, has a severe brittleness problem and has low tensile strength. Few studies have been carried out to reduce the brittleness of concrete [6, 8, 9]. The incorporation of fibers into plain concrete is an effective way to reduce intrinsic brittleness, particularly in OPC. The most typical method involves adding a precise amount of fibers to reinforced binder components. Fibers were put into concrete to improve flexural strength, ductility, toughness, and durability, among other properties. The fibers were used to bridge the crack in the matrix while also transferring the applied load to the matrix. As a result, concrete reinforced with fibers exhibits superior post-crack behavior than plain concrete [7]. The insertion of fibers can improve flexural and tensile strength by increasing impact resistance. However, various factors influence concrete improvement, including binder strength, fiber volume, surface bonding properties, fiber shape, and percentage of fiber addition [10-12].

Materials and Methodology

Fly ash was used as an alumino-silicate source material, and an alkaline activator consisting of sodium silicate (Na_2SiO_3) and sodium hydroxide was used to activate the geopolymer concrete (NaOH). Fly ash is a waste product produced by the combustion of coal as a raw source. Na_2SiO_3, also known as water glass, has a SiO/N_2O ratio of 3.2. Meanwhile, a certain ratio of NaOH pellet to distilled water was used to make a NaOH solution with a concentration of 12 M.

X-ray fluorescence (XRF) was used to determine the chemical composition of the raw material fly ash, and the kind of fly ash was justified based on the results. The ratio of $Na_2SiO_3/NaOH$ employed in alkaline activator synthesis was 2.5. Meanwhile, the proportion of fly ash to alkaline activator utilized in this study was 2.0.

River sand with a maximum size of 4 mm was used as fine aggregate in this project. Meanwhile, the coarse aggregate employed in this study had a size of 16 mm. Hooked steel fiber and polypropylene (PP) fiber of good grade were employed in this study. Steel and PP fiber were each cut to a thickness of 2 mm. Steel and PP fiber were added to the geopolymer concrete in percentages ranging from 0% to 2% of the total volume. Table 1 shows the mix design of the steel-PP fiber reinforced geopolymer concrete.

Composite Materials: SEAJCCM2024   Materials Research Forum LLC
Materials Research Proceedings 56 (2025) 93-100   https://doi.org/10.21741/9781644903636-10

Table 1. Mix design of steel-PP fiber reinforced geopolymer concrete.

Sample	Steel fiber (%)	PP fiber (%)	Fly ash (kg/m³)	Coarse aggregate (kg/m³)	Fine aggregate (kg/m³)	NaOH 12 M (kg/m³)	Na₂SiO₃ (kg/m³)
S0P0	0.0	0.0	496.8	745.2	496.8	71.0	177.4
S0P2	0.0	2.0	496.8	745.2	496.8	71.0	177.4
S0.5P1.5	0.5	1.5	496.8	745.2	496.8	71.0	177.4
S1P1	1.0	1.0	496.8	745.2	496.8	71.0	177.4
S1.5P0.5	1.5	0.5	496.8	745.2	496.8	71.0	177.4
S2P0	2.0	0.0	496.8	745.2	496.8	71.0	177.4

For density, water absorption, and compression tests, samples were cast in cubic sizes of 100 mm X 100 mm X 100 mm. After 24 hours in the mold, all samples were cured at room temperature. After 14 days, the samples were examined for density, workability, and compression. The workability of geopolymer concrete was determined by a slump test by following ASTM C143. Equation 1 (Eq. 1) was used to calculate sample density. Equation 2 (Eq. 2) was used to determine water absorption. Meanwhile, a 1000 kN Universal Testing Machine (UTM) was used for compression testing. This experimental procedure following the British standard which is a method to determine the compressive strength of concrete cubes BS 1881-116.

$$Density \left(\frac{kg}{m^3}\right) = \frac{W_d}{W_s - W_i} x1000 \tag{1}$$

$$Water\ absorption\ (\%) = \frac{W_s - W_d}{W_d} x100\% \tag{2}$$

Where:

W_d = oven dry weight of specimen (kg)
W_s = saturated weight of specimen (kg)
W_i = immersed weight of specimen (kg)

Results and Discussions
Table 2 shows the results of the chemical composition of fly ash that was analyzed using XRF. Table 2 shows that the CaO level was 18.10 %, indicating that this fly ash is classed as Class F according to ASTM C618. According to ASTM C618, any fly ash with a CaO concentration of less than 20% is classified as Class F [13]. According to XRF results, the total $Al_2O_3 + SiO_2 + Fe_2O_3$ of this fly ash is 72.98 %. The ratio of silica to alumina (SiO_2/Al_2O_3) in fly ash is also in the optimum range of 2.64.

Table 2. Chemical composition of fly ash by using XRF.

Chemical Composition	Percentage (%)
SiO₂	38.80
Fe2O₃	19.48
CaO	18.10
Al2O₃	14.70
MgO	3.30
K2O	1.79
SO₃	1.50
BaO	0.27
MnO	0.16
SrO	0.11
L.O.I	2.41

Composite Materials: SEAJCCM2024   Materials Research Forum LLC
Materials Research Proceedings 56 (2025) 93-100   https://doi.org/10.21741/9781644903636-10

Fig. 1 shows a downward trend, where the slump value decreases as the percentage of the volume of steel increases, whereas the percentage of PP decreases. Based on the results, the highest slump value for the fly ash-based geopolymer concrete is 100.67 mm with no fiber additions. Samples with the addition of 0.0 % steel and 2.0 % PP, 0.5 % steel, and 1.5 % PP obtained slump values of 92.87 mm and 84.33 mm respectively. Other than that, samples with fiber addition of 1.0 % steel and 1.0 % PP, and 1.5 % steel and 0.5 % PP yield 80.11 mm and 72.69 mm of slump values respectively. The additions of both fibers seem to lower the slump of geopolymer concrete compared to unreinforced concrete. This is supported by previous studies stating that adding fibers will reduce the slump of normal concrete [14-15].

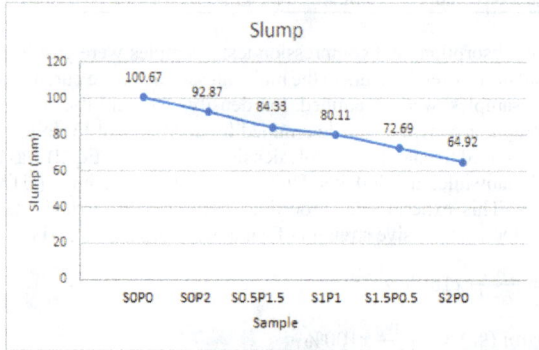

Fig. 1. The effect of steel and PP fibers on the slump.

Sample with fiber addition of 2.0 % steel and 0.0 % PP achieved the lowest slump value which is 64.92 mm. The lower slump is due to the ability of steel fibers to form a network structure in concrete, which prevents mixture segregation and flow. Because of the high content and huge surface area of the fibers, they will absorb more geopolymer paste to wrap around them, and the increased viscosity of the mixture will cause slump loss. Besides, the trend shows that steel fibers have a higher impact on lowering the slump compared to the addition of PP fibers. This is due to the higher flexibility of PP compared to steel fibers (high stiffness) in which the PP fibers can bend more easily in the presence of interlocking conditions. Hence, the resistance to the flow of fresh concrete is lower with the addition of PP fibers. Other studies also show a similar trend in which they stated the higher the stiffness of the fiber the less the slump value [15-16].

Based on Fig. 2, the highest density is obtained by the geopolymer concrete containing 2 % steel and 0 % which is 2460 kg/m^3, followed by the second highest which is 1.5 % steel and 0.5 % PP geopolymer concrete with a density of 2451 kg/m^3. The lowest density is obtained by the geopolymer concrete containing 0 % steel and 2 % PP which is 2317 kg/m^3, followed by the second lowest which is the control sample or the geopolymer concrete containing 0 % addition of fiber with the density of 2320 kg/m^3.

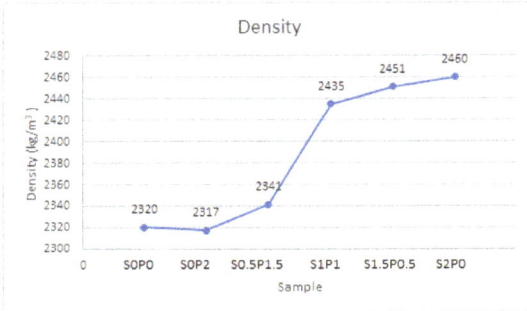

Fig. 2. The density of steel-PP geopolymer concrete.

It can be concluded that the steel fiber contributes greatly towards the density of the geopolymer concrete compared to the PP fiber. This is because the PP fiber has a very low density with 946 kg/m^3, whereas the density of steel fiber is very high with 7850 kg/m^3. It was mentioned in previous research that the density of geopolymer concrete does get higher as the addition of steel fiber increases [17].

As shown in Fig. 3, geopolymer concrete containing 0 % steel and 2 % PP obtained the highest water absorption percentage with 7.11 %. On the other hand, geopolymer concrete with 2 % steel and 0 % PP obtained the lowest water absorption percentage at 5.71 %. Water absorption is usually related to the density of the geopolymer concrete, where a higher density of geopolymer concrete results in lower water absorption. This is due to the lower amount of porosity that existed in the geopolymer concrete sample. Hence, the weight of the geopolymer concrete increases as the geopolymer matrix fills up most of the voids, thus increasing the density of the geopolymer concrete. This is supported by previous study where they conclude water absorption reduces with an increase of density for geopolymer samples [18]. The increase in the addition of steel fiber to the geopolymer concrete will result in a decrease in the absorption of water into the sample and will cause the bonding stronger thus increasing the compressive strength. It is proven that steel and PP fiber has a significant impact on the void structures of the geopolymer material.

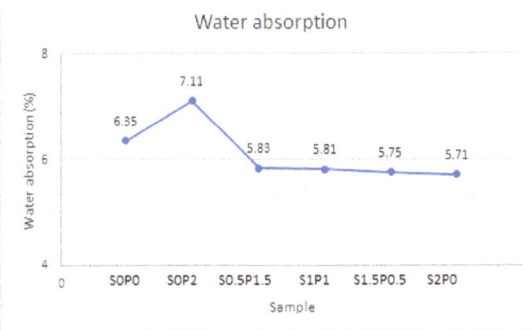

Fig. 3. Water absorption percentage of steel-PP geopolymer concrete.

Based on the graph plotted in Fig. 4, the highest value of compressive strength of geopolymer concrete is 49.31 MPa with fiber addition of 2 % steel and 0 % PP, followed by the sample with 1.5 % steel and 0.5 % PP which obtained 47.04 MPa. The increase in compressive strength

Composite Materials: SEAJCCM2024   Materials Research Forum LLC
Materials Research Proceedings 56 (2025) 93-100   https://doi.org/10.21741/9781644903636-10

observed with the addition of fibers can be attributed to the significant contribution of steel fibers, which play a key role in enhancing the post-cracking behavior of the material. These fibers work by improving the concrete's ability to resist further cracking and distribute stresses more effectively after the initial failure. In their research, Cuenca et al. [19] demonstrated that steel fibers are particularly effective at bridging cracks in ordinary Portland cement (OPC) concrete, provided that the total fiber content is sufficient to perform this function, thereby improving the overall structural integrity and durability of the concrete.

The lowest value of compressive strength of geopolymer concrete is 24.45 MPa with fiber addition of 0 % steel and 2 % fiber, which is lower compared to the control sample which has no fiber additions. This may be due to the chemical nature of PP makes it hydrophobic concerning the cementitious matrix, resulting in diminished cement bonding and a detrimental impact on its dispersion in the matrix. This is stated by a previous study where PP fibers claimed to have a hydrophobic surface lead to no existence of physical-chemical adhesion bonding between PP fibers and cement binders during the mixing process [20]. Hence, the mechanical performance including compressive strength will decrease. Besides that, due to the balling effect while mixing, PPF may have reduced the strength. Steel fiber is usually used as reinforcement to improve the performance of geopolymer concrete. However, the sample with fiber addition of 0.5 % steel and 1.5 % PP obtained the second lowest compressive strength, which was 29.21 MPa. The higher young modulus of steel fiber compared to PP fiber used in the geopolymer concrete could be the reason for this trend in which the higher the steel fibers content the higher the compressive strength.

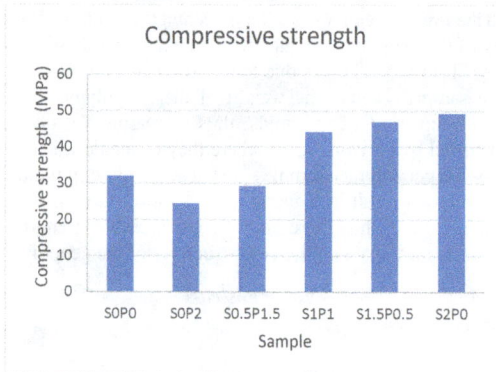

Fig. 4. Compressive strength of steel-PP geopolymer concrete.

Under axial stresses, fissures in the microstructure of geopolymer concrete and fibers are observed to limit the formation and growth of cracks by generating clamping forces at crack tips. They take on some of the stress that happens in the geopolymer matrix and transfer the rest to stable geopolymer matrix regions. It decreased the fracture-tip stress concentration, preventing the crack from propagating forward and changing its route. The fracture was blunted, blocked, and even diverted, allowing the geopolymer concrete cubes with a higher steel fiber percentage to sustain more compressive stress, improving their compressive strength over the geopolymer concrete cubes with PP fiber.

Conclusion

The amount of steel and PP fiber used has significantly impacted the mechanical and physical properties of geopolymer concrete. The workability decreases as the steel fiber increases and the

Composite Materials: SEAJCCM2024   Materials Research Forum LLC
Materials Research Proceedings 56 (2025) 93-100   https://doi.org/10.21741/9781644903636-10

PP fiber decreases. The higher amount of PP will lead to a higher water absorption percentage and a higher steel percentage result in the decrement of water absorption percentage. Besides that, the properties of geopolymer concrete are also affected by its density. In this case, a higher amount of PP fiber in the geopolymer concrete will greatly decrease its density of the geopolymer concrete. However, the density will increase as the amount of steel fiber in geopolymer concrete increases. Besides, the steel-PP composite fiber reinforced geopolymer concrete achieved its optimum mechanical properties through the geopolymer concrete with the addition of 2 % steel and 0 % PP. Geopolymer concrete with the addition of 2 % steel and 0 % PP has obtained the highest value in compressive strength. In terms of compressive strength, it may be said that the addition of 2% steel fiber and 0% PP fiber to geopolymer concrete is the best among the others. Suggestions for future studies that may lead to investigating the long-term durability of geopolymer concrete reinforced with varying steel and PP fiber content under different environmental conditions (e.g., exposure to moisture, temperature variations, and chemical attacks) would be beneficial. Besides, conducting microstructural investigations (e.g., SEM or X-ray diffraction) to understand the bonding mechanisms between the fibers and the geopolymer matrix will help in optimizing fiber reinforcement strategies for improved performance.

Acknowledgments
The authors would like to express their sincere thanks to the Ministry of Higher Education of the Malaysian Government for giving the Fundamental Research Grant Scheme (FRGS) to support this research with a grant number of 9003-00936.

References
[1] P. Duxson, A. Fernández-Jiménez, J.L. Provis, G.C. Lukey, A. Palomo, J.S. van Deventer, Geopolymer technology: the current state of the art, Journal of materials science 42 (2007) 2917-2933. https://doi.org/10.1007/s10853-006-0637-z

[2] D.L. Kong, J.G. Sanjayan, K. Sagoe-Crentsil, Comparative performance of geopolymers made with metakaolin and fly ash after exposure to elevated temperatures, Cement and concrete research 37 (2007) 1583-1589. https://doi.org/10.1016/j.cemconres.2007.08.021

[3] K.H. Yang, J.K. Song, K.I. Song, Assessment of CO2 reduction of alkali-activated concrete, Journal of Cleaner Production 39 (2013) 265-272. https://doi.org/10.1016/j.jclepro.2012.08.001

[4] N. Lloyd, V. Rangan, Geopolymer concrete with fly ash, in: Proceedings of the Second International Conference on sustainable construction Materials and Technologies, 2010, 1493-1504. UWM Center for By-Products Utilization.

[5] E. Hewayde, M. Nehdi, E. Allouche, G. Nakhla, Effect of geopolymer cement on microstructure, compressive strength and sulphuric acid resistance of concrete, Magazine of Concrete Research 58 (2006) 321-331. https://doi.org/10.1680/macr.2006.58.5.321

[6] M. Soutsos, A.P. Boyle, R. Vinai, A. Hadjierakleous, S.J. Barnett, Factors influencing the compressive strength of fly ash based geopolymers, Construction and Building Materials 110 (2016) 355-368. https://doi.org/10.1016/j.conbuildmat.2015.11.045

[7] C.D. Atiş, E.B. Görür, O.K.A.N. Karahan, C. Bilim, S.ER. H.A.N. İlkentapar, E. Luga, Very high strength (120 MPa) class F fly ash geopolymer mortar activated at different NaOH amount, heat curing temperature and heat curing duration, Construction and building materials 96 (2015) 673-678. https://doi.org/10.1016/j.conbuildmat.2015.08.089

[8] E. Hewayde, M. Nehdi, E. Allouche, G. Nakhla, Effect of geopolymer cement on microstructure, compressive strength and sulphuric acid resistance of concrete, Magazine of Concrete Research 58 (2006) 321-331. https://doi.org/10.1680/macr.2006.58.5.321

Composite Materials: SEAJCCM2024   Materials Research Forum LLC
Materials Research Proceedings 56 (2025) 93-100   https://doi.org/10.21741/9781644903636-10

[9] X. Ma, Z. Zhang, A. Wang, The transition of fly ash-based geopolymer gels into ordered structures and the effect on the compressive strength, Construction and Building Materials 104 (2016) 25-33. https://doi.org/10.1016/j.conbuildmat.2015.12.049

[10] H.T. Wang, L.C. Wang, Experimental study on static and dynamic mechanical properties of steel fiber reinforced lightweight aggregate concrete, Construction and Building Materials 38 (2013) 1146-1151. https://doi.org/10.1016/j.conbuildmat.2012.09.016

[11] P.S. Song, S. Hwang, Mechanical properties of high-strength steel fiber-reinforced concrete, Construction and Building Materials 18 (2004) 669-673. https://doi.org/10.1016/j.conbuildmat.2004.04.027

[12] M. Valle, O. Buyukozturk, Behavior of fiber reinforced high-strength concrete under direct shear, Materials Journal 90 (1993) 122-133. https://doi.org/10.14359/4006

[13] ASTM Standard C618, 2022. Standard Specification for Coal Fly Ash and Raw or Calcined Natural Pozzolan for Use in Concrete.

[14] A. Chajec, Ł. Sadowski, The Effect of Steel and Polypropylene Fibers on the Properties of Horizontally Formed Concrete, Materials 13 (2020) 5827. https://doi.org/10.3390/ma13245827

[15] V. Guerini, A. Conforti, G. Plizzari, S. Kawashima, Influence of Steel and Macro-Synthetic Fibers on Concrete Properties, Fibers 6 (2018) 47. https://doi.org/10.3390/fib6030047

[16] A.D. De Figueiredo, M.R. Ceccato, Workability analysis of steel fiber reinforced concrete using slump and Ve-Be test, Mater. Res. 18 (2015) 1284–1290. https://doi.org/10.1590/1516-1439.022915

[17] M.M.A.B. Abdullah, M.F.M. Tahir, M.A.F.M.A. Tajudin, J.J. Ekaputri, R. Bayuaji, N.A.M. Khatim, Study on the geopolymer concrete properties reinforced with hooked steel fiber, IOP Conference Series: Materials Science and Engineering 267 (2017) 012014. https://doi.org/10.1088/1757-899X/267/1/012014

[18] E.M. Kumar, K. Ramamurthy, Influence of production on the strength, density and water absorption of aerated geopolymer paste and mortar using Class F fly ash, Construction and Building Materials 156 (2017) 1137-1149. https://doi.org/10.1016/j.conbuildmat.2017.08.153

[19] E. Cuenca, J. Echegaray-Oviedo, P. Serna, Influence of concrete matrix and type of fiber on the shear behavior of self-compacting fiber reinforced concrete beams, Composites Part B: Engineering 75 (2015) 135-147. https://doi.org/10.1016/j.compositesb.2015.01.037

[20] L. Akand, M. Yang, X. Wang, Effectiveness of chemical treatment on polypropylene fibers as reinforcement in pervious concrete, Construction and Building Materials 163 (2018) 32-39. https://doi.org/10.1016/j.conbuildmat.2017.12.068

Composite Materials: SEAJCCM2024
Materials Research Proceedings 56 (2025) 101-107

Materials Research Forum LLC
https://doi.org/10.21741/9781644903636-11

Physical and Mechanical Properties of Steel-Polypropylene Composites Reinforced OPC Concrete

Meor Ahmad FARIS[1,2,a *], Warid Wazien AHMAD ZAILANI[3,b],
Muhammad Fahem MOHD TAHIR[2,4,c], Mohammad Firdaus ABU HASHIM[1,2,d],
M.N.A. UDA[1,5,e], Mohd Ikram RAMLI[6,f], Mohammed Izzuddeen MOHD YAZID[7,g]

[1]Faculty of Mechanical Engineering Technology, Universiti Malaysia Perlis (UniMAP), Pauh Putra 026000 Arau, Perlis, Malaysia

[2]Center of Excellence Geopolymer and Green Technology (CEGeoGTech), Universiti Malaysia Perlis, 01000 Kangar, Perlis, Malaysia

[3]School of Civil Engineering, College of Engineering, Universiti Teknologi MARA (UiTM), 40450 Shah Alam, Malaysia

[4]Faculty of Chemical Engineering Technology, University Malaysia Perlis, 01000 Kangar, Perlis, Malaysia

[5]Institute of Nano Electronic Engineering, Universiti Malaysia Perlis, 01000 Kangar, Perlis, Malaysia

[6]School of Engineering, Universiti of Wollongong Malaysia, Jalan Kontraktor U1/14, Seksyen U1, 40150 Shah Alam, Malaysia

[7]Faculty of Innovative Design & Technology, Universiti Sultan Zainal Abidin, 21030 Kuala Terengganu, Malaysia

[a]meorfaris@unimap.edu.my, [b]waridwazien@uitm.edu.my, [c]faheem@unimap.edu.my, [d]firdaushashim@unimap.edu.my, [e]nuraiman@uitm.edu.my, [f]ikram.r@uow.edu.my, [g]mohdizzuddeen@unisza.edu.my

Keywords: Geopolymer, Steel Fiber, Polypropylene Fiber, Reinforced Concrete, Hybrid Fibers

Abstract. Fiber reinforced concrete (FRC) consists of a cement matrix with randomly dispersed fibers, enhancing its overall durability compared to conventional concrete. Ordinary Portland Cement (OPC) concrete faces significant challenges, and incorporating varying percentages of steel and polypropylene fibers may influence its physical and mechanical properties. This study aims to produce OPC-based reinforced concrete with composite fibers specifically, steel and polypropylene to determine the optimal fiber ratio for enhanced performance. The advantages of this research lie in identifying the ideal hybrid fiber ratio for producing high-performance concrete, along with determining the most suitable materials for specific conditions. Concrete samples with different fiber ratios will be tested to evaluate their physical and mechanical characteristics ranging from 0 % - to 2 %. Slump, density, and water absorption tests were conducted for physical analysis. Meanwhile, compressive strength is conducted for mechanical analysis. The slump test shows a reduction of workability when fibers are added compared to unreinforced concrete and shows a significant effect on the addition of steel compared to PP fibers. Density and water absorption show an effect of the addition of steel and PP fibers. The density of samples with additions of various fibers ranges from 2216 kg/m³ to 2451 kg/m³ which is within the ranges of normal concrete. Besides, the addition of hybrid fibers shows a range of water absorption 6.37 % (S1.5PP0.5) to 8.61 % (unreinforced concrete). Meanwhile, the compression test shows an

Content from this work may be used under the terms of the Creative Commons Attribution 3.0 license. Any further distribution of this work must maintain attribution to the author(s) and the title of the work, journal citation and DOI. Published under license by Materials Research Forum LLC.

Composite Materials: SEAJCCM2024   Materials Research Forum LLC
Materials Research Proceedings 56 (2025) 101-107   https://doi.org/10.21741/9781644903636-11

improvement in the sample by adding hybrid steel-PP fibers. The findings indicate that a mixture containing 1.5% steel and 0.5% polypropylene fibers offers the highest performance.

Introduction

Concrete is known for its excellent compressive strength, but its tensile strength is considerably lower which is only about one-tenth that of steel. This makes enhancing the tensile properties of concrete a crucial area of study, especially for applications that need high energy absorption such as bridges, tunnels, dams, and stadiums. These building structures require superior mechanical properties such as high compressive, tensile, and flexural strength. Besides, the physical properties of concrete including the ability to flow smoothly around dense reinforcement or adapt to complex shapes, as well as considerations related to economic and environmental sustainability are important factors that need to be considered [1].

Fiber Reinforced Concrete (FRC) is a concrete that utilizes the potential of fibers that are embedded into the concrete system to enhance the mechanical properties of normal concrete and reduce the possibility of crack propagation under certain loads. The type, shape, and dimensions of the fibers including length (l), diameter (d), and aspect ratio (slenderness l/d) are crucial factors that influence the effectiveness of FRC [2]. Additionally, the material properties of the fibers such as chemical composition and physical characteristics play a significant role in determining the behavior of FRC. Meanwhile, increasing fiber content can positively impact the physical and mechanical properties of concrete and simply adding more fibers does not necessarily guarantee improved performance [3].

The effectiveness of fiber reinforcement largely depends on the bonding force between the fibers and the concrete matrix, as well as the length-to-diameter ratio of the fibers [4]. According to elastic fracture mechanics, fiber spacing plays a key role in determining tensile strength, especially at a specific fiber volume content [5]. A well-chosen blend of fibers can address inherent weaknesses in concrete, enhancing its hardness, reducing stress concentration at crack tips, and preventing the formation and propagation of internal cracks. However, traditional models may not fully capture the complexities of fiber reinforcement mechanisms [5], suggesting the need for further research to optimize fiber types, ratios, and integration techniques for various applications.

A recent study shows the intention of researchers to hybridize various types of fibers to improve their mechanical properties as well as to reduce cost [6, 7]. This paper aims to find the potential of a combination of hybrid fibers of steel and polypropylene that purposely enhances the mechanical properties of OPC concrete while at least maintaining the physical properties at an accepted level. The study has the potential to explore related to the percentages of addition between two combinations of different fibers (hybrid) that affect the performance and durability of normal concrete. The purpose is to ensure higher energy absorption to reduce brittleness and higher reliability that offers practical insight for meeting a variety of structural needs.

Materials and Methodology

In this study, straight steel fibers (SF) were used to enhance the performance of OPC concrete. The physical properties of the straight steel fibers used in this research are detailed in Table 1. Steel fibers were added to the OPC concrete mix based on percentage by volume by the following ratios of 0.0%, 0.5%, 1.0%, 1.5%, and 2.0%. Table 1 outlines the physical properties of the steel fibers utilized in the experiments.

Composite Materials: SEAJCCM2024
Materials Research Proceedings 56 (2025) 101-107

Materials Research Forum LLC
https://doi.org/10.21741/9781644903636-11

Table 1. Physical properties of steel fibers.

No.	Property	Values
1	Diameter	0.75 mm
2	Length of fiber	20 mm
3	Appearance	Bright in clean wire
4	Average aspect ratio	80
5	Deformation	Straight
6	Tensile strength	1050 MPa
7	Modulus of elasticity	200 GPa
8	Specific gravity	7.8

Polypropylene fiber is also added to the concrete by the percentage volume same as the steel fiber. The size of the coarse aggregate used in this experiment is ranging 5 mm to 20 mm and the specific gravity of 2.62. The fine aggregate that has been used is river sand the size is 4 mm and below. The steel fibers are prepared and sorted into a single piece before being added to the OPC concrete mixture. There are two sizes of mold are prepared to be casting the concrete. The samples used in the slump, density, water absorption, and compression tests were 100 mm x 100 mm x 100 mm.

The mixing started with the preparation and accurate measurement of materials, including OPC, steel fibers, polypropylene fibers, and aggregates. These dry components were first thoroughly mixed to ensure a uniform distribution of fibers. After achieving a consistent blend, water was gradually added to the mixture. The mixture was then allowed to solidify before undergoing additional mixing to ensure a homogeneous consistency.

Once the mixture was properly mixed, it was poured into pre-prepared molds to form samples. The freshly cast samples were kept at a controlled temperature of 27°C for the initial 24 hours to facilitate the initial setting. After 24 hours, the specimens were carefully demolded and transferred to a curing environment. The samples were cured at room temperature for 28 days to ensure optimal hydration and proper hardening of the concrete before any testing was conducted.

The density and water absorption of the product were calculated using Eq. 1 and Eq. 2, respectively. Each sample's weight is weighted and documented in terms of immersed weight W_i, saturated weight W_s, and oven dry weight W_d. The density test was calculated using the average change in mass of each sample in these three terms. The density will be calculated by using the formula shown in Eq. 1.

$$Density \left(\frac{kg}{m^3}\right) = \frac{W_d}{W_s - W_i} x1000 \tag{1}$$

$$Water\ absorption\ (\%) = \frac{W_s - W_d}{W_d} x100\% \tag{2}$$

Where:

W_d = oven dry weight of specimen (kg)
W_s = saturated weight of specimen (kg)
W_i = immersed weight of specimen (kg)

The slump test was conducted according to ASTM C143. The concrete mix is poured in three layers into a slump mold set on a flat surface, with each layer tamped 25 times using a rod to ensure uniformity. Each layer is tamped throughout the depth of the mold. If the concrete settles below the top edge, the mold is topped off. After the top layer is levelled, any excess at the base is cleaned off. The slump value is determined by measuring the height difference between the top of the mold and the slumped concrete. Then, the compressive strength test was performed using a SHIMADZU Universal Testing Machine (UTM). The size of the specimen that we used in this testing session is 100 mm x 100 mm x 100 mm. All results were recorded and analyzed.

Results and Discussions

The result of the slump test for the OPC concrete by additions of the steel and polypropylene fibers is shown in Fig. 1. The highest value of the slump test was recorded which is 97.3 mm and this sample is not mixed with any fibers. The value of the slump test starts to drop dramatically when the steel fiber is added to the mixing ratio which the value is 16.3 mm. From the graph in Fig. 1, the value of the slump test is increasing due to the decrease in steel and the increase in polypropylene. 34.6 mm slump was recorded from the sample with a ratio of steel 1.5 %, polypropylene 0.5 % and for the mixing ratio S1PP1, the value of slump is 46.5 mm.

Slump tests were conducted to evaluate the workability of the OPC mixture. Based on various situations and applications, different concrete structures of different applications need different workability which range from 10 mm to 220 mm [8]. Based on the graph, the value of slumps is within the ranges to be used in normal concrete for various applications. As shown in Fig. 1, the slump value decreases with an increase in steel fiber content. This suggests that the presence of fibers creates resistance to the free flow of the OPC mixture. Consequently, the addition of steel fibers reduces the workability of the concrete, particularly because the proportion of steel fibers is higher compared to polypropylene fibers in the mixture.

Fig. 2 shows the density properties of the composite reinforced OPC concrete which the sample 28 days curing. From the graph in Fig. 2, the highest density of the OPC concrete is with inclusions 1.5 % steel mix with 0.5 % polypropylene which is 2451 kg/m^3 and the second highest is 2400 kg/m^3 where the ratio is using 2 % steel not including polypropylene. The density for the control sample is 2316 kg/m^3 and the lowest density was recorded on the sample with the ratio of 2 % polypropylene not mixing with steel.

As illustrated in Fig. 2, the density of OPC concrete decreases with a reduction in the percentage of steel fibers in the mix. This is because the inclusion of fibers increases the overall density of the OPC concrete. Higher density in OPC concrete typically results in greater compressive strength, while lower density concrete tends to have increased porosity and higher water absorption. This is supported by a previous study where they stated the density was inversely proportional to porosity for a sample that has been tested involved in normal concrete [9]. Porosity can negatively affect the concrete's strength. To minimize porosity, it is crucial to prevent air bubbles from forming during the mixing process, as these bubbles can burst and create pores that compromise the concrete's integrity.

The highest percentage of water absorption as in Fig. 3 is 8.61 % from the control sample which no fiber is added to the concrete. The percentage of water absorption decreases due to the addition of the fibers into the concrete and the percentage starts to drop from 8.61 % to 7.11 % when the 2.0 % of steel fiber is added. The lowest percentage of water absorption was recorded from the sample with the ratio 1.5 steel 0.5 % polypropylene where the value of the percentage of water absorption is 6.37 %.

The results show that the percentage of water absorption has started to rise from 6.37 % to 7.72 % of water absorption from the ratio S1.5PP0.5 to the S1PP1. The percentage of water absorption increases due to the decreasing volume of steel fiber respectively. Meanwhile, the existing amount of PP seems to increase the water absorption hence increasing the porosity. The concrete is better when the percentage of water absorption is low. The water absorption is higher when the open porosity of the concrete is also higher. Because of the porosity, cracks may form in the sample, lowering the compressive strength. The open pore reflects the ability of air and water to pass through the OPC concrete mixture and make it permeable. This is supported by previous studies

Composite Materials: SEAJCCM2024   Materials Research Forum LLC
Materials Research Proceedings 56 (2025) 101-107   https://doi.org/10.21741/9781644903636-11

where they stated the density and porosity of samples will affect the permeability of concrete [9, 10]. The higher the porosity the higher the ability of concrete to absorb water.

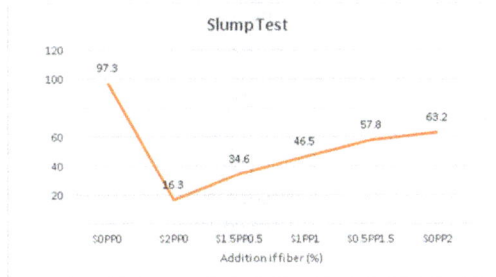

Fig. 1. Slump of OPC with the addition of various combinations of hybrid steel-PP fibers.

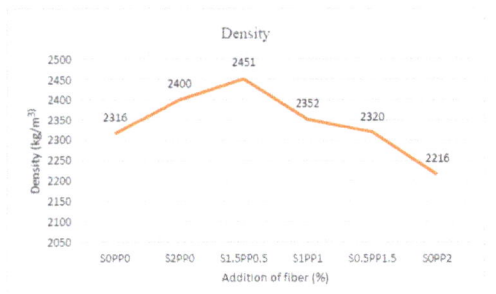

Fig. 2. The density of OPC concrete with additions of various combinations of hybrid steel-PP fibers.

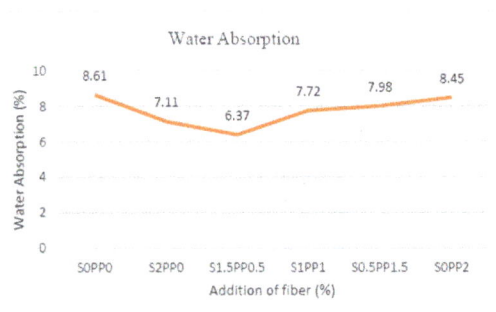

Fig. 3. Water absorption of OPC concrete with the addition of various combinations of hybrid steel-PP fibers.

Fig. 4 presents the results of the compressive strength test conducted on OPC concrete samples after 28 days of curing, with varying inclusions of steel and polypropylene fibers. The lowest recorded compressive strength, 10.2 MPa, was observed in the control sample with no fiber additions. The highest compressive strength which is 25.2 MPa is achieved in the sample containing 1.5% steel and 0.5% polypropylene fibers. The second-highest compressive strength is 19.1 MPa which is found in the sample with 2% steel and no polypropylene fibers.

The results indicate that increasing the proportion of steel fibers generally enhances the compressive strength of OPC concrete. This clearly shows the contribution of steel fibers to provide a bridging effect to stop the cracks propagations when forces are added is higher compared to additions of PP fibers. Therefore, the optimal ratio for compressive strength was 1.5% steel and 0.5% polypropylene. Adding more steel fibers beyond this ratio results in a decrease in compressive strength. The compressive strength results indicate that the presence of PP will reduce the performance. However, the combination of both steel and polypropylene fibers significantly enhances the concrete's strength. The inclusion of fibers not only improves the mechanical properties of the concrete but also mitigates the effects of cracking. Specifically, the steel and polypropylene fibers effectively inhibit crack propagation, thereby enhancing the overall durability of the concrete [11].

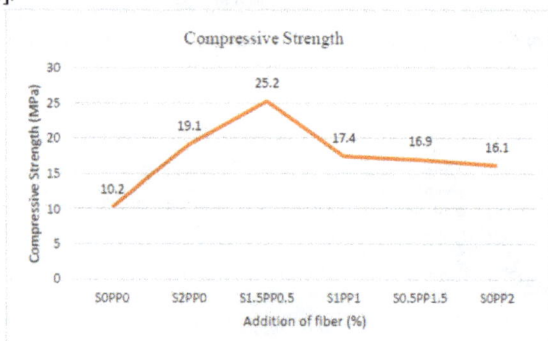

Fig. 4. Compressive strength of OPC concrete with various percentages addition of hybrid steel-PP fibers.

Conclusion
In conclusion, the study demonstrates that the incorporation of steel-polypropylene (PP) fibers significantly enhances the mechanical and physical properties of Ordinary Portland Cement (OPC) concrete. Increasing the PP fiber content reduces the workability of OPC concrete. As indicated by slump testing, it contributes to several desirable characteristics. Specifically, higher PP fiber content leads to a reduction in water absorption which leads to enhancing the concrete's durability. However, increasing fiber content will lead to a decrease in the concrete's density but still in the acceptance range for normal concrete. The findings highlight that the optimal addition of 1.5 % steel with 0.5 % PP fibers not only boosts the compressive strength of OPC concrete but also improves its physical properties which is for water absorption while maintaining the density and slump in an acceptance range for various construction applications. This study provides valuable insights for constructors, enabling them to better assess the performance of steel-PP fiber reinforced concrete before application. Consequently, the results offer a clear pathway to minimizing flaws in OPC concrete and enhancing the overall properties of fiber-reinforced concrete for improved structural applications.

Acknowledgments
The authors would like to express their sincere thanks to the Ministry of Higher Education of the Malaysian Government for giving the Fundamental Research Grant Scheme (FRGS) to support this research with a grant number of 9003-00936.

Composite Materials: SEAJCCM2024   Materials Research Forum LLC
Materials Research Proceedings 56 (2025) 101-107   https://doi.org/10.21741/9781644903636-11

References

[1] J. Blazy, R. Blazy, Polypropylene fiber reinforced concrete and its application in creating architectural forms of public spaces, Case Studies in Construction Materials 14 (2021) e00549. https://doi.org/10.1016/j.cscm.2021.e00549

[2] N.R. Iyer, An overview of cementitious construction materials, New materials in civil engineering (2020) 1-64. https://doi.org/10.1016/B978-0-12-818961-0.00001-6

[3] K.R. Kumar, G. Shyamala, A. Adesina, Structural performance of corroded reinforced concrete beams made with fiber-reinforced self-compacting concrete, Structures 32 (2021) 1145-1155. https://doi.org/10.1016/j.istruc.2021.03.079

[4] P.A. Nagar, N. Gupta, K. Kishore, A.K. Parashar, Coupled effect of B. Sphaericus bacteria and calcined clay mineral on OPC concrete, Materials Today: Proceedings 44 (2021) 113-117. https://doi.org/10.1016/j.matpr.2020.08.029

[5] X.K. Li, L. Sun, Y.Y. Zhou, S.B. Zhao, A review of stee-polypropylene hybrid fiber reinforced concrete, Applied Mechanics and Materials 238 (2012) 26-32. https://doi.org/10.4028/www.scientific.net/AMM.238.26

[6] S. Das, M.H.R. Sobuz, V.W. Tam, A.S.M. Akid, N.M. Sutan, F.M. Rahman, Effects of incorporating hybrid fibres on rheological and mechanical properties of fibre reinforced concrete, Construction and Building Materials 262 (2020) 120561. https://doi.org/10.1016/j.conbuildmat.2020.120561

[7] M. Mastali, A. Dalvand, A.R. Sattarifard, Z. Abdollahnejad, M.J.C.P.B.E. Illikainen, Characterization and optimization of hardened properties of self-consolidating concrete incorporating recycled steel, industrial steel, polypropylene and hybrid fibers, Composites Part B: Engineering 151 (2018) 186-200. https://doi.org/10.1016/j.compositesb.2018.06.021

[8] S.H. Kosmatka, W.C. Panarese, B. Kerkhoff, Design and control of concrete mixtures, 5420 (2002) 60077-1083. Skokie, IL: Portland cement association

[9] A. Akkaya, İ.H. Çağatay, Investigation of the density, porosity, and permeability properties of pervious concrete with different methods, Construction and Building Materials 294 (2021) 123539. https://doi.org/10.1016/j.conbuildmat.2021.123539

[10] V. Revilla-Cuesta, F. Faleschini, C. Pellegrino, M. Skaf, V. Ortega-López, Water transport and porosity trends of concrete containing integral additions of raw-crushed wind-turbine blade, Developments in the Built Environment 17 (2024) 100374. https://doi.org/10.1016/j.dibe.2024.100374

[11] A. Jameran, I.S. Ibrahim, S.H.S. Yazan, S.N.A. Rahim, Mechanical properties of steel-polypropylene fibre reinforced concrete under elevated temperature, Procedia Engineering 125 (2015) 818-824. https://doi.org/10.1016/j.proeng.2015.11.146

Composite Materials: SEAJCCM2024

Materials Research Proceedings 56 (2025) 108-116

Materials Research Forum LLC

https://doi.org/10.21741/9781644903636-12

The Impact of Surfactant-Driven Emulsion Droplet Size on Lead (Pb) Recovery Efficiency in Emulsion Liquid Membrane Systems

N.H. IBRAHIM[1,a], S.N. ZAILANI[1,b] *, N.A. ZAINOL[1,c],
Subash C.B. GOPINATH[1,2,d], M.N.A. UDA[3,4,5,e], Ahmad Radi WAN YAAKUB[1,f],
M.N.A. UDA[6,g] and U. HASHIM[6,h]

[1]Faculty of Chemical Engineering & Technology, Universiti Malaysia Perlis, 02600 Arau, Perlis, Malaysia

[2]Institute of Nano Electronic Engineering, Universiti Malaysia Perlis, 01000 Kangar, Perlis, Malaysia

[3]Faculty of Mechanical Engineering & Technology, Universiti Malaysia Perlis, 02600 Arau, Perlis Malaysia

[4]Institute of Nano Electronic Engineering, Universiti Malaysia Perlis, 01000 Kangar, Perlis, Malaysia

[5]Centre of Excellence for Biomass Utilization, Universiti Malaysia Perlis, 02600 Arau, Perlis, Malaysia

[6]Faculty of Engineering, Universiti Malaysia Sabah, 88400 Kota Kinabalu, Sabah, Malaysia

[a]hulwaniibrahim@gmail.com, [b]sitinazrah@unimap.edu.my, [c]aineezainol@unimap.edu.my,
[d]subash@unimap.edu.my, [e]nuraiman@unimap.edu.my, [f]ahmadradi@unimap.edu.my,
[g]nurafnan@ums.edu.my, [h]uda@ums.edu.my

Keywords: Lead, Surfactant, Removal and Recovery Process, Emulsion Liquid Membrane

Abstract. The emulsion liquid membrane (ELM) extraction technology offers significant perspectives into the extraction and recovery of lead (Pb) from aqueous solutions. In this study, ELM stability was investigated using ELM components consist of bis(2-ethylhexyl) phosphate (D2EHPA) as an extractant, kerosene as diluent, nitric acid (HNO_3) as stripping agent and polysorbate 80 (Tween 80) as surfactant. These components were prepared and formulated at 2000 rpm and 5 minutes emulsification time. The size of emulsion droplet and the effect of surfactant concentration on the breakage/swelling effect for ELM stability and extraction efficiency were studied. The results show that, at 0.5 M D2EHPA, 0.5 M HNO_3 and 5 % w/v Tween 80 was enough to perform stable emulsion with 1.5 μm droplet size and the highest extraction efficiency of Pb removal was 96.16 %. ELM approach has significant potential to facilitate extraction process efficiently.

Introduction

The rising concentrations of lead (Pb) ions in industrial wastewater, which might reach 200–500 mg/L in the environment, seriously jeopardize people's health over long terms [1]. These hazards are severe, once Pb enters the food chain, it can accumulate in the human body at high concentrations. Ingesting metals beyond permitted levels may lead to serious health disorders. Therefore, Pb needs to be removed in aqueous form to sustain the community's health and environment. Furthermore, the recovery of Pb from liquid waste plays a vital role in environmental sustainability through utilization processes. The Malaysian Sewage and Industrial Effluent Discharge Standards have been implemented in Malaysia, which specify that the permissible level of Pb should not exceed 0.5 mg/L [2]. The objective of implementing restrictions on wastewater

Content from this work may be used under the terms of the Creative Commons Attribution 3.0 license. Any further distribution of this work must maintain attribution to the author(s) and the title of the work, journal citation and DOI. Published under license by Materials Research Forum LLC.

Composite Materials: SEAJCCM2024   Materials Research Forum LLC
Materials Research Proceedings 56 (2025) 108-116   https://doi.org/10.21741/9781644903636-12

was to mitigate the potential risk of human and environmental exposure to hazardous substances [3]. Even small amounts of these substances, especially Pb, are highly toxic. However, the management and prevention of heavy metal pollution continue to be the crucial issues in a global scale, despite the fact that it has been a problem for a long time [4].

In addition, a great number of research has concentrated on the elimination of Pb, while a smaller number of studies have addressed the recovery process of Pb. Adsorption, ion exchange, electrochemical techniques and others were employed to remove Pb from inorganic effluent [5-9]. However, these methods have several significant drawbacks, incomplete elimination, demand for a significant amount of energy and result in the development of harmful sludge [10]. Moreover, this traditional method is laborious and demands a substantial quantity of chemicals for its execution. Therefore, this research introduces a comprehensive technique involving the emulsion liquid membrane (ELM) system, which offers valuable insights into both the removal and recovery of Pb from aqueous solutions. ELM presents a promising approach with its integrated extraction and stripping process in a single step, which is considered efficient for Pb recovery. The formulation of ELM components comprises selecting the extractant, surfactant, diluent and stripping agents focusing on achieving the highest extraction and stripping rates of Pb in aqueous solutions [11].

The development of ELM components is personalized specifically for targeted solutes, with Pb being the focus of this study. A milky white emulsion is prepared using highly selective and uniquely designed extractants, surfactants, and diluents impregnated with the stripping phase, along with optimized parameters studied [12]. ELM is a complex dispersion system that involves dispersing an emulsion containing organic and aqueous internal phases into the aqueous external phase containing solutes. Fig.1 shows the fundamental principle of a three-phase in ELM system. The solutes that are present in the external phase are transferred to the membrane phase, where they are subjected to a chemical reaction with the stripping agent that is present in the internal phase. This reaction takes place in the membrane phase. Furthermore, this material is able to infiltrate into the other phases of the membrane, nevertheless, it may able to stay completely contained inside the internal phase [13].

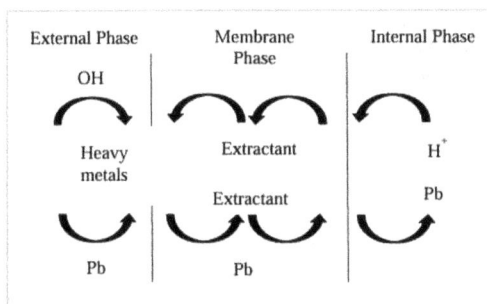

Fig. 1. The process of Pb ions transportation in ELM.

Maintaining the extractant and stripping rate capabilities of the device requires the ELM to be stable. This is a very important requirement. An extremely stable emulsion may generate difficulties such as settling, which in turn leads to a loss in the efficacy of extraction and stripping. This is even though a stable emulsion can provide high extraction and stripping rates. However, an excessively stable emulsion may cause problems. Several factors were considered to investigate

Composite Materials: SEAJCCM2024   Materials Research Forum LLC
Materials Research Proceedings 56 (2025) 108-116   https://doi.org/10.21741/9781644903636-12

the stability of the ELM. All factors were evaluated, including surfactant concentration, particle size generated by the emulsion and the preparation method.

Experimental Analysis

Reagents

ELM consists of an organic phase, an internal phase (also known as a stripping agent) and an external phase. The organic phase comprises of D2EHPA (97% purity, ACROS Organic, Belgium), kerosene (99% purity, Sigma-Aldrich (M), Darmstadt, Germany) and Tween 80 (98% purity, Sigma-Aldrich (M), Darmstadt, Germany). The internal phase consisted of HNO_3 (65% purity, Sigma-Aldrich (M), Darmstadt, Germany). The external phase, on the other hand, consisted of lead nitrate $(PbNO_3)_2$ (99% purity, Sigma-Aldrich (M), Darmstadt, Germany).

Table 1. Displays the specific experimental parameters employed for the formulation of ELM system.

Parameter	Condition
Extractant	0.5 M D2EHPA
Diluent	kerosene
Surfactant	Tween 80
Concentration of surfactant	3, 5 and 7 % w/v
Stripping agent	0.5 M HNO_3
Initial Pb Concentration	10 ppm
Treat Ratio	1:1
Stirring speed	250 rpm
Emulsification time	5 min

ELM preparation

D2EHPA was dissolved in kerosene and Tween 80 as surfactant was added at the concentration of 3, 5 and 7% w/v respectively into the solution to form an organic phase. The emulsion was prepared by emulsifying formulated organic phase with stripping phase (internal phase). An equal volume of 5 mL organic and 5 mL stripping phase was stirred continuously using motor driven homogenizer with stirring speed 2000 rpm and 5 minutes stirring time to obtain a stable white milky emulsion liquid membrane. The experiments were carried out in triplicates. ELM preparation process was shown in Fig. 2.

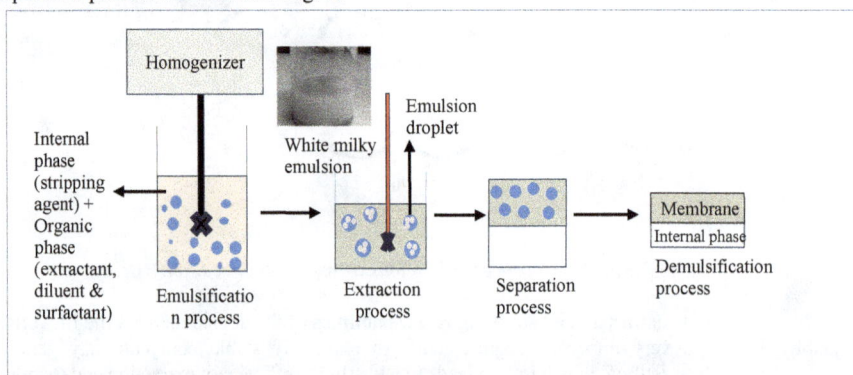

Fig. 2. ELM preparation process.

Composite Materials: SEAJCCM2024   Materials Research Forum LLC
Materials Research Proceedings 56 (2025) 108-116   https://doi.org/10.21741/9781644903636-12

The average mean diameter of emulsion was measured using an optical microscope (Olympus, BX51, UK) as shown in Fig. 3. One drop of emulsion was placed on a glass slide and by using a microscope equipped with a camera, a few photos were captured. The droplet diameter can be expressed as Sauter diameter (d32) which represents the average surface diameter in Eq. 1 as reported in the previous study [14].

$$d_{32} = \frac{\sum_i n_i d_{i2}}{\sum_i n_i d_{i3}} = 6\frac{V}{A} \tag{1}$$

Where, $n_i d_i$ are the number and diameter of droplets (μm), while V and A are the total volume (m^3) and area (cm^2) of the dispersed phase, respectively.

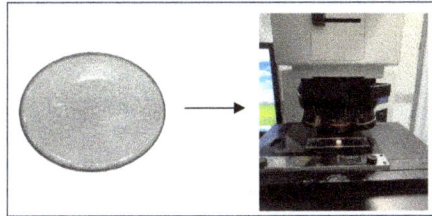

Fig. 3. Emulsion droplet size.

ELM stability study

The prepared emulsion was dispersed into 10 mL of 10 ppm Pb solution at the treat ratio 1:1 (volume of emulsion to Pb solution). An aqueous Pb solution was served as the external feed phase. This mixture was homogenized at 250 rpm for 5 minutes. Upon completing the extraction process, samples (emulsion and aqueous Pb solution) were quickly introduced into a separation funnel and left for phase separation. The aqueous phase was analyzed by using AAS. Then, the volume of emulsion was recorded. The swelling and breakage effect occurred when the emulsion was not stable enough to resist the external force during the extraction process. The increase in the volume of emulsion shows the increment of water in the internal phase, while decrease in the volume of emulsion shows the breakage effect. The experiments were carried out in triplicates. The percentage emulsion stability was calculated using Eq. 2 as reported in the previous study [15].

$$\% \text{ Stability} = \frac{V_i - V_i^0}{V_i^0} \times 100\% \tag{2}$$

Where, V_i is the volume (mL) of the internal phase after extraction and V_i^0 is the initial volume (mL) of the internal phase.

Results and Discussion

Emulsion droplet size

Emulsion droplet size is a critical factor that directly impacts the stability and effectiveness of extraction and recovery process in ELM. Smaller droplets exhibit higher resistance to breaking and allow for faster extraction. On the other hand, bigger droplets lead to poor ELM stability and lower efficiency in extraction [16]. Fig. 4 shows that, increasing surfactant concentration from 3 to 7 % w/v, decreases the emulsion droplet size from 4 μm to 0.98 μm. The surface area that is available for mass transfer between the feed phase (the phase that contains the substance that is being targeted) and the internal phase of the emulsion is increased as the droplets of the emulsion are smaller. This enhances the rate at which the target molecules able to diffuse into the emulsion droplet, making the separation process faster and more efficient.

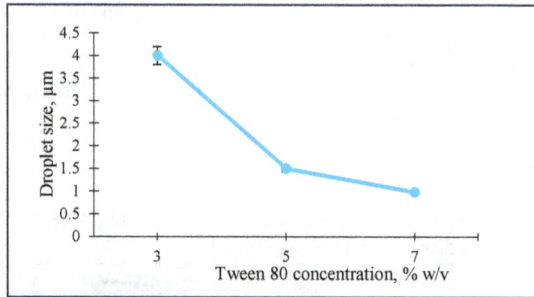

Fig. 4. Effect of tween 80 concentration on the emulsion droplet size (Experimental condition: D2EHPA=0.5 M, HNO3=0.5 M, diluent=kerosene, tween 80=3, 5 and 7 % w/v, homogenizer speed=2000 rpm, emulsifying time=5 minutes).

It was observed by Yu et al. [17] that emulsions that comprise droplets with a size ranging from 0.3 to 10 μm (preferably 0.8 to 3 μm) demonstrate good extraction rates and a considerable stability of emulsion-like materials (ELM). There is a possibility that difficulties associated with settling and de-emulsification may arise during the third stage of the ELM process. This is more evidence that an excessive amount of emulsion stability, which is characterized by decreased droplet size, is not desirable. Thus, 5 % w/v Tween 80 emulsion appears to be the most stable with 1.5 μm droplet size as depicted in Fig. 5. This concentration provides sufficient surfactant coverage to stabilize the droplets without leading to excess aggregation or droplet merging, resulting in a well-dispersed and controlled emulsion. Too much surfactant (7 % w/v) can disrupt the balance, leading to over-saturation of the droplet interface and potentially destabilizing the emulsion.

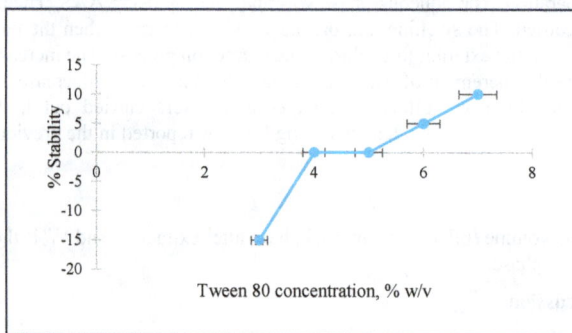

Fig. 5. Emulsion droplet size at different Tween 80 concentrations.

Effect of surfactant concentration

Stable emulsion is crucial for effective extraction and recovery of Pb. A well-formed emulsion generates a large surface area, enhancing mass transfer between the aqueous and organic phases [18], which is the key for efficient metal transport. Appropriate surfactant concentration is essential to form a stable emulsion. Fig. 6 demonstrates that stability improved with an increase in surfactant concentration. It can be clearly seen that, from 4 to 5 % w/v of Tween 80, the emulsion remains stable. Tween 80 molecules adsorb well onto the oil droplets, forming a strong protective barrier that prevents coalescence. With no breakage or swelling (0 %), 5 % Tween 80 (% w/v) appears to

Composite Materials: SEAJCCM2024   Materials Research Forum LLC
Materials Research Proceedings 56 (2025) 108-116   https://doi.org/10.21741/9781644903636-12

be the optimal for emulsion stability. Tween 80 concentration is high enough to create a stable interface, reducing the chance of droplet fusion. The interfacial tension is minimized and the droplets remain dispersed. These results align with those of previous research [19]. At low concentrations (3 % w/v), Tween 80 is insufficient to stabilize the emulsion, resulting in breakage. As concentration increases, Tween 80 improves stability by lowering interfacial tension, providing steric hindrance and forming a hydration barrier around droplets. This process makes the emulsion droplets less likely to coalesce, thus enhancing stability. At even higher concentrations (7 % w/v), Tween 80 can attract water molecules to the surfactant layer around each droplet, causing slight swelling [20]. This means that at 3 % w/v, indicating the emulsion breakage while at 7 % w/v indicating some degree of swelling to the emulsion.

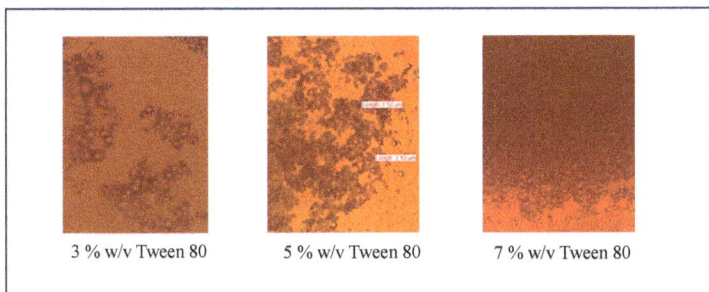

| 3 % w/v Tween 80 | 5 % w/v Tween 80 | 7 % w/v Tween 80 |

Fig. 6. Effect of tween 80 concentration on the stability of emulsion (Experimental condition: D2EHPA=0.5 M, HNO3=0.5 M, diluent=kerosene, Tween 80= 3, 5 and 7 % w/v, homogenizer speed=2000 rpm, emulsifying time=5 minutes, agitation speed=250 rpm, extraction time=5 minutes).

Tween 80 has several properties that make it ideal for producing stable emulsions. It is widely used as a surfactant because of its unique structure, chemical stability and compatibility with both aqueous and organic phases. The key properties that make Tween 80 suitable for stable emulsion production is it has a high Hydrophilic-Lipophilic Balance (HLB) value of around 15, indicating that it is highly hydrophilic [21]. In this case, the optimal HLB range for the W/O/W ELM system is typically between 8 to 16. Given that Tween 80 itself has an HLB of 15, it is ideal for this type of emulsion. High HLB value makes Tween 80 particularly effective for creating oil-in-water (O/W) emulsions, where the oil phase is dispersed within the continuous aqueous phase allowing Tween 80 to effectively stabilize the interface between oil and water. Then, successfully produce stable emulsion in W/O/W ELM system using the formulation of D2EHPA, kerosene and HNO3.

Extraction study
The extraction of Pb depends on the quality of the emulsion. Once the emulsion is dispersed in Pb aqueous solution, the organic droplets are spread throughout the aqueous solution. The extractant within the organic droplets is selective for Pb ions. These ions in the aqueous phase will react with D2EHPA and being extracted by a cation exchange process between Pb^{2+} in the aqueous phase and H^+ derived from D2EHPA in the organic phase. Eq. 3 depicts the extraction process of Pb ions, illustrating the typical interaction between Pb ions and D2EHPA molecules via its representation.

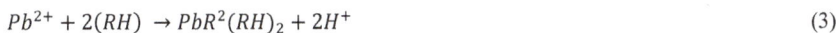

$$Pb^{2+} + 2(RH) \rightarrow PbR^2(RH)_2 + 2H^+ \tag{3}$$

Based on Fig. 7, the bar graph depicts that, at 5 % surfactant concentration, extraction efficiency for Pb is the highest at 96.16 %. At 3 and 7 % surfactant concentration, extraction efficiency is relatively low which is 19.21 % and 17.06 %, respectively. At 5 % Tween 80, a good emulsion provides larger surface area of the interface between the aqueous and organic phases that leads to more efficient on extraction of Pb [22]. Hence, this suggests that 3 and 7 % w/v surfactant may not provide adequate Pb removal. Tween 80 forms micelles or emulsions that interact with Pb^{2+} ions, transferring them from the aqueous phase into the organic phase through ion-dipole interactions or encapsulation within micelles [23]. The transfer of Pb from the aqueous to the organic phase is efficiently enhanced by Tween 80 at an appropriate concentration (5 % w/v) via the creation of stable micelles or emulsions that interact with and encapsulate Pb^{2+} ions [24].

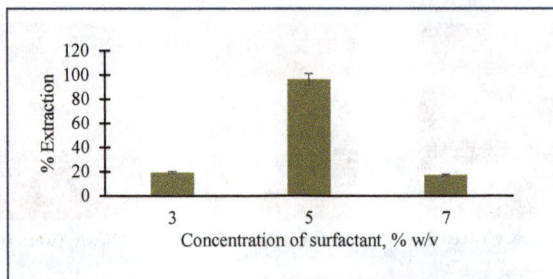

Fig. 7. Extraction of Pb at different surfactant concentration (Experimental condition: D2EHPA=0.5 M, HNO₃=0.5 M, diluent=kerosene, Tween 80=3, 5 and 7 % w/v, initial Pb concentration=10 ppm, agitation speed=250 rpm, extraction time=5 minute).

Conclusion

This study investigates the stability and effectiveness of the ELM process using Tween 80 as the surfactant, D2EHPA as the extractant, kerosene as the diluent, and HNO_3 as the stripping agent. This formulation successfully produced a stable emulsion for the extraction and recovery of Pb from an aqueous solution. The emulsion demonstrates excellent stability at a concentration of 5 % w/v Tween 80, with a droplet size of 1.5 μm. The dimensions of emulsion droplets decrease as the surfactant concentration increases from 3 to 7% w/v, respectively. Thus, the best stable emulsion was obtained at a concentration of 5 % w/v of Tween 80. The concentration of surfactant significantly affects emulsion droplet size. As a result, the optimal extraction efficiency for Pb removal is 96.16 %. The concentration of the surfactant at an optimal level of 5 % w/v is substantially more effective in facilitating the transfer of Pb from the aqueous phase to the organic phase. This is in contrast to the concentrations of 3 and 7 % w/v. When it comes to Pb extraction and recovery, the ELM process is a procedure that is both successful and promising.

Acknowledgements

The authors gratefully acknowledge the research funding provided by the Ministry of Higher Education Malaysia (MOHE) and the Fundamental Research Grant Scheme (FRGS) Ref: FRGS/1/2019/TK10/UNIMAP/02/20 (9003-00778). The authors also extend their appreciation to the Faculty of Chemical Engineering and Technology and the Institute Nano Electronic Engineering (INEE) at Universiti Malaysia Perlis, as well as the technical staff for their invaluable support in facilitating this research.

Composite Materials: SEAJCCM2024
Materials Research Proceedings 56 (2025) 108-116

Materials Research Forum LLC
https://doi.org/10.21741/9781644903636-12

References

[1] M. de J. Soria-Aguilar, A. Martínez-Luévanos, M.A. Sánchez-Castillo, F.R. Carrillo-Pedroza, N. Toro, V.M. Narváez-García, Removal of Pb(II) from aqueous solutions by using steelmaking industry wastes: Effect of blast furnace dust's chemical composition,Arab. J. Chem. 14 (2021) 4. https://doi.org/10.1016/j.arabjc.2021.103061

[2] T.N.B.T. Ibrahim, F. Othman, N.Z. Mahmood, Baseline Study of Heavy Metal Pollution in a Tropical River in a Developing Country, Sains Malaysiana, 49 (2020) 729–742. https://doi.org/10.17576/jsm-2020-4904-02

[3] K. Yunus, M.A. Zuraidah, A. John, A review on the accumulation of heavy metals in coastal sediment of Peninsular Malaysia, Ecofeminism Clim. Chang. 1 (2020) 21–35. https://doi.org/10.1108/efcc-03-2020-0003

[4] E. Quitério, C. Grosso, R. Ferraz, C. Delerue-Matos, C. Soares, A Critical Comparison of the Advanced Extraction Techniques Applied to Obtain Health-Promoting Compounds from Seaweeds, Mar. Drugs 20 (2022) 1–40. https://doi.org/10.3390/md20110677

[5] S.Y. Tsai, C.K. Chang, P.Y. Wei, S.Y. Huang, M. Gavahian, S.P. Santoso, C.W. Hsieh, Effective Removal of Different Heavy Metals Ion (Cu, Pb, and Cd) from Aqueous Solutions by Various Molecular Weight and Salt Types of Poly-γ-Glutamic Acid, Molecules 29 (2024) 1054. https://doi.org/10.3390/molecules29051054

[6] A. Pohl, Removal of Heavy Metal Ions from Water and Wastewaters by Sulfur-Containing Precipitation Agents, Water. Air. Soil Pollut. 231 (2020) 503. https://doi.org/10.1007/s11270-020-04863-w

[7] A. Lalmi, K.E. Bouhidel, B. Sahraoui, and C. el H. Anfif, Removal of lead from polluted waters using ion exchange resin with Ca(NO3)2 for elution, Hydrometallurgy 178 (2017) 287–293. https://doi.org/10.1016/j.hydromet.2018.05.009

[8] L. Gurreri, A. Tamburini, A. Cipollina, and G. Micale, Electrodialysis applications in wastewater treatment for environmental protection and resources recovery: A systematic review on progress and perspectives, Membranes (Basel) 10 (2020) 1–93. https://doi.org/10.3390/membranes10070146

[9] G. Crini, E. Lichtfouse, Advantages and disadvantages of techniques used for wastewater treatment, Environ. Chem. Lett. 17 (2019) 145–155. https://doi.org/10.1007/s10311-018-0785-9

[10] M.A. Agoro, A.O. Adeniji, M.A. Adefisoye, O.O. Okoh, Heavy metals in wastewater and sewage sludge from selected municipal treatment plants in eastern cape province, south africa, Water (Switzerland) 12 (2020) 2746. https://doi.org/10.3390/w12102746

[11] N. Qudus, A. Kusumastuti, S. Anis, A.L. Ahmad, Emulsion liquid membrane for lead removal: Intensified low shear extraction, Int. J. Innov. Learn. 25 (2019) 285–295. https://doi.org/10.1504/IJIL.2019.098887

[12] S. Lakshminarayana, A. Shilpika, N. Kumar, P. Kranthireddy, K. Deepak, S. Usha, M. Vijayakumar, An Experimental Study on Partial Replacement of Cement in Concrete with Sugarcane Bagasse Ash Using Magnetized Water, Lect. Notes Civ. Eng., 352 (2024) 143–150. https://doi.org/10.1007/978-981-99-2676-3_12

[13] A. Kusumastuti, S. Anis, A.L. Ahmad, B.S. Ooi, M.M.H. Shah Buddin, Emulsion liquid membrane for heavy metals removal: Emulsion breaking study, J. Teknol. 82 (2020) 51–57. https://doi.org/10.11113/jt.v82.14539

Composite Materials: SEAJCCM2024   Materials Research Forum LLC
Materials Research Proceedings 56 (2025) 108-116   https://doi.org/10.21741/9781644903636-12

[14] K. Vhora, G. Janiga, H. Lorenz, A. Seidel-Morgenstern, M.F. Gutierrez, P. Schulze, Comparative Study of Droplet Diameter Distribution: Insights from Experimental Imaging and Computational Fluid Dynamics Simulations, Appl. Sci., 14 (2024) 1824. https://doi.org/10.3390/app14051824

[15] H.M. Salman, A.A. Mohammed, Extraction of lead ions from aqueous solution by co-stabilization mechanisms of magnetic Fe2O3 particles and nonionic surfactants in emulsion liquid membrane, Colloids Surfaces A Physicochem. Eng. Asp. 568 (2019) 301–310. https://doi.org/10.1016/j.colsurfa.2019.02.018

[16] H.M. Azwatul, M.N.A. Uda, S.C. Gopinath, Z.A. Arsat, F. Abdullah, M.F.A. Muttalib, M.K.R. Hashim, U. Hashim, M. Isa, M.N. Afnan Uda, A. R.W. Yaakub, N.H. Ibrahim, N.A. Parmin, T. Adam, Synthesis and characterization of silver nanoparticle using sewage algal bloom extract using visual parameter analysis, Mater. Today Proc. (2023) https://doi.org/10.1016/j.matpr.2023.01.004

[17] S. Yu, J. Zhang, S. Li, Z. Chen, Y. Wang, Mass Transfer and Droplet Behaviors in Liquid-Liquid Extraction Process Based on Multi-Scale Perspective: A Review, Separations 10 (2023) 264. https://doi.org/10.3390/separations10040264

[18] M. Khadivi, V. Javanbakht, Emulsion ionic liquid membrane using edible paraffin oil for lead removal from aqueous solutions, J. Mol. Liq. 319 (2020) 114137. https://doi.org/10.1016/j.molliq.2020.114137

[19] S.S. Kulkarni, V.A. Juvekar, S. Mukhopadhyay, Intensification of emulsion liquid membrane extraction of uranium(VI) by replacing nitric acid with sodium nitrate solution, 125 (2018) 18-26. https://doi.org/10.1016/j.cep.2017.12.021

[20] M. Zamouche, H. Tahraoui, Z. Laggoun, S. Mechati, R. Chemchmi, M.I. Kanjal, A. Amrane, A. Hadadi, L. Mouni, Optimization and Prediction of Stability of Emulsified Liquid Membrane (ELM): Artificial Neural Network, Processes 11 (2023) 364. https://doi.org/10.3390/pr11020364

[21] N.F.M. Noah, N. Othman, N. Jusoh, I.N.S. Kahar, S.S. Suliman, Succinic Acid Recovery and Enhancement of Emulsion Liquid Membrane Stability using Synergist Aliquat 336/TOA/Palm Oil System Assisted with Nanoparticle, Arab. J. Sci. Eng. 48 (2023) 15777–15792. https://doi.org/10.1007/s13369-023-07616-z

[22] B. Bera, R. Khazal, K. Schroën, Coalescence dynamics in oil-in-water emulsions at elevated temperatures, Sci. Rep. 11 (2021) 10990. https://doi.org/10.1038/s41598-021-89919-5

[23] M.G.A. Kassem, A.M. Ahmed, H.H. Abdel-rahman, H. E. Moustafa, Use of Span 80 and Tween 80 for blending gasoline and alcohol in spark ignition engines, Energy Reports 5 (2019) 221–230. https://doi.org/10.1016/j.egyr.2019.01.009

[24] G. Tartaro, H. Mateos, D. Schirone, R. Angelico, G. Palazzo, Microemulsion microstructure(s): A tutorial review, Nanomaterials 10 (2020) 1657. https://doi.org/10.3390/nano10091657

Composite Materials: SEAJCCM2024
Materials Research Proceedings 56 (2025) 117-123

Materials Research Forum LLC
https://doi.org/10.21741/9781644903636-13

Effect of Silicon and Mechanical Alloying on Microstructure and Grain Evolution in CuAl and CuAlSi Alloys

Nurlyana Izyan MOHD ALI[1,a], Nur Izzati MUHAMMAD NADZRI[1,b *],
Arif Anuar MUHD SALLEH[1,c], M.N.A. UDA[2,3,4,d] and Sudha JOSEPH[5,e]

[1]Center of Excellence Geopolymer and Green Technology (CEGeoGTech), Universiti Malaysia Perlis, 01000 Kangar, Perlis, Malaysia

[2]Institute of Nano Electronic Engineering, Universiti Malaysia Perlis, 01000 Kangar, Perlis, Malaysia

[3]Faculty of Mechanical Engineering Technology, Universiti Malaysia Perlis, 02600 Arau, Perlis, Malaysia

[4]Centre of Excellence for Biomass Utilization, Universiti Malaysia Perlis, 02600 Arau, Perlis, Malaysia

[5]Department of Materials Engineering, Cambridge Institute of Technology, Bengaluru, Karnataka 560036, India

[a]nurlyana@studentmail.unimap.edu.my, [b]izzatinadzri@unimap.edu.my,
[c]arifanuar@unimap.edu.my, [d]nuraiman@unimap.edu.my,
[e]sudhajoseph.cccir@cambridge.edu.in

Keywords: Powder Metallurgy, Mechanical Alloying, Grain Coalescence

Abstract. This research examines the influence of Si addition in CuAl alloy, and the effect of mechanical alloying on the microstructure and elemental distribution of CuAl and CuAlSi alloys, fabricated using powder metallurgy. SEM, micro-XRF, and EDX analyses were conducted to study morphological changes, compositional uniformity, and microstructural evolution. Mechanical alloying was found to introduce strain and defects, which, coupled with sintering, facilitated grain boundary movement and coarsening. Micro-XRF and EDX results showed improved elemental distribution in ball-milled and sintered samples compared to as-is alloys. In CuAlSi alloys, the equimolar composition supported uniform element distribution, though oxidation during sintering reduced the metallic Cu content. Sintering, especially following mechanical alloying, improved the microstructure by increasing grain compactness and refining elemental distribution. These modifications enhance the alloys' mechanical properties, making them suitable for high-performance applications.

Introduction

In recent years, the development of advanced materials has increasingly focused on optimizing alloy compositions and processing techniques to enhance mechanical properties and performance [1]. Copper and its alloys are widely used in many industrial applications such as marine, energy, and transportation. Copper-aluminium (CuAl) alloys are widely used in various industries due to their excellent thermal conductivity, corrosion resistance, and mechanical strength for applications such as heat exchangers in thermal power plants, and brazing alloy for diamond grits [2-6]. In Cu-Al alloys, aluminium acts as a solid solution strengthener, improving both hardness and yield strength (YS) [7]. Additionally, the incorporation of microalloying elements such as Mn, silicon (Si), Ni, Zr, Cr, and Ag can enhance mechanical properties through precipitation strengthening by forming second-phase particles [8]. Among these microalloying elements, Si and magnesium (Mg) exhibit sufficient solid solubility in both copper and aluminium, significantly influencing precipitation behavior and kinetics in Cu-Al alloys [9]. Magnesium, when present at

Content from this work may be used under the terms of the Creative Commons Attribution 3.0 license. Any further distribution of this work must maintain attribution to the author(s) and the title of the work, journal citation and DOI. Published under license by Materials Research Forum LLC.

Composite Materials: SEAJCCM2024
Materials Research Proceedings 56 (2025) 117-123

Materials Research Forum LLC
https://doi.org/10.21741/9781644903636-13

concentrations of 0.3–1%, enhances oxidation resistance, thermal conductivity, and electrical conductivity. Meanwhile, the addition of silicon improves the fluidity of the molten metal and minimizes the formation of intermetallic phases or intermetallic compounds (IMCs) [10-11]. Conventional fabrication methods often result in inhomogeneous elemental distribution and irregular grain growth, necessitating the exploration of advanced processing techniques such as mechanical alloying and sintering to address these limitations.

This study investigates the effect of (1) Si addition and (2) ball milling on the microstructure, and the elemental distribution of CuAl alloy fabricated using the powder metallurgy method was investigated. Advanced characterization methods such as SEM, μ-XRF, and EDX are used to analyse microstructural changes, and compositional uniformity during sintering. The selection of Cu, Al, and Si is due to their complementary properties which are thermal stability, and mechanical strength which make these alloys highly versatile for structural and functional applications.

Materials and Methodology

High purity (99.9%) Cu, Al, and Si powders were used to prepare samples with different alloy compositions, shown in Table 1, and were fabricated by using the powder metallurgy technique. Two sets of samples were prepared where one set underwent mechanical alloying, while the other set was left unprocessed. The powder forms as a raw material having a meshing size < 200 μm. Mechanical alloying was performed using a planetary ball mill (Fritsch Pulverisette Planetary Ball Mill) in the presence of stainless-steel balls with a ball-to-powder ratio of 10:1. Powder was milled at 120 rpm for 2 h. The powder mixture was compacted in a stainless-steel mold under uni-axial direction by using a Specac-15-Ton Manual Hydraulic Press Machine under 400 MPa load for 3 min. After cold compaction, samples were acquired in cylindrical shape having a diameter ≈ of 10 mm and a height of approximately 2 mm. The homogenous powder was sintered by a conventional sintering process. The temperature profile is shown in Fig. 1. A muffle furnace was used to sinter the alloys at 600 °C. The sample heating rate was 5 °C/min throughout the complete sintering cycle and furnace cooled. Preparations of samples were carried out by grinding, and polishing. Scanning electron microscopy (SEM, JEOL JSM-6460LA) was used to examine the surface morphology of the alloy samples. The elements distributions were analyzed by using a synchrotron micro-X-ray fluorescence spectrometer (μ-XRF Synchrotron Light Research Institute (SLRI) BL6b beamline located in Thailand).

Table 1. Compositions of CuAlSix alloys in weight percentage.

Elements (wt.%)	Cu	Al	Si
CuAl	0.25	0.25	0.00
CuAlSi	0.166	0.166	0.166

Results and Discussion

Microstructural Characteristics of CuAl Alloy

Fig. 1. illustrates the micrographs of the CuAl alloy, capturing various processing stages: (a) non-ball-milled and non-sintered CuAl powders, and (b) CuAl alloy after 3 hours of ball milling and sintering. Fig. 1(c-f) provides the elemental distributions of Cu and Al, showcasing (c) as-is Cu powder, (d) sintered Cu powder, (e) as-is Al powder, and (f) sintered Al powder. The darker regions in Fig. 1(a-b) show Al-rich zones, while the brighter regions correspond to Cu-rich zones.

The microstructure in Fig. 1(b) reveals coarser and more elongated grains compared to the finer grains in the unprocessed powder shown in Figure 1(a). The particle size of Al powder is 17.4 μm and 10.5μm for Cu powder zones. In comparison to Al-rich zone in milled and sintered which is 18μm and 20μm for Cu-rich regions. This is due to powders undergoing mechanical deformation during milling where the impact forces cause the particles to flatten and elongate which results in a change in their shape and size distribution [12]. In comparison, the unprocessed alloy retains

Composite Materials: SEAJCCM2024   Materials Research Forum LLC
Materials Research Proceedings 56 (2025) 117-123   https://doi.org/10.21741/9781644903636-13

finer grains and exhibits more pronounced compositional segregation due to the absence of mechanical activation and thermal treatment. This is supported by research on Mg-Gd-Zn alloy where sintering temperature affects the grains. As the sintering temperature increased from 430°C to 470°C, the grain sizes increased from 4.6μm to 5.8μm. This is due to enhanced atomic mobility and boundary migration during thermal treatment [13].

The micro-XRF elemental maps for Cu and Al, shown in Fig. 1(c-f), provide insights into the distribution of these elements in both green and sintered bodies. The distribution becomes more uniform and dispersed after the ball milling and sintering processes. The colour bar indicates elemental concentration, with reddish hues representing higher concentrations and bluish hues indicating lower concentrations. In the sintered samples, the increased reddish and greenish hues observed in Fig. 1(d-f) suggest higher concentrations of Cu and Al compared to the as-is samples. The elemental maps confirm that the sintering process significantly enhances compositional uniformity by homogenizing the distribution of elements across the sample. Mechanical alloying followed by sintering improves the microstructure's compactness and ensures a more uniform distribution of elements, contributing to better material properties in the final product.

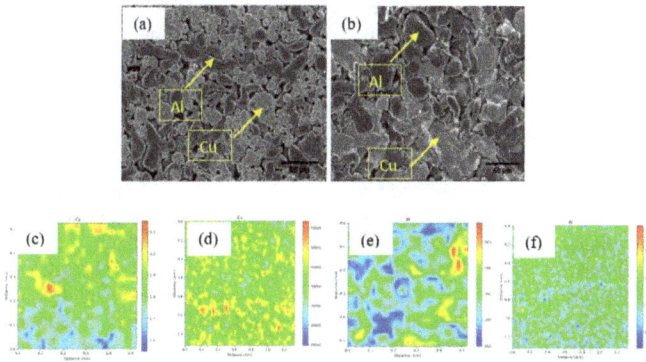

Fig. 1. Microstructure of CuAl alloy where (a) non-ball milli and non-sintered CuAl powders, (b) 3 hours ball milled and sintered CuAl alloy. Elemental distributions of Cu and Al elements where (c) as-is Cu powder, (d) sintered Cu powder, (e) as-is Al powder, and (f) sintered Al powder.

Microstructural Characteristics of CuAlSi alloy

Fig. 2 shows the microstructures of sintered CuAlSi alloy, comparing 2(a) a sample without ball milling and 2(b) a sample subjected to 3 hours of ball milling. Similar to the CuAl alloy, the sintered CuAlSi alloy exhibits larger grains after ball milling. This is due to the ball mill introducing significant mechanical energy into powder particles, leading to defects. These defects can enhance the atomic mobility during sintering, where it led to coarsening on grains. During sintering, the defects created by ball milling promote grain migration and facilitate atomic diffusion. This leads to grain coalescence, where smaller grains merge into larger ones, resulting in coarser particles. The stored energy from mechanical alloying reduces the energy barrier for grains, enabling faster grain migration and densification during sintering [14]. In contrast, non-ball-milled samples retain their finer grain structure due to the absence of these driving forces, limiting atomic diffusion. The study by Kulecki et al. highlights that mechanical activation through ball milling affects particle morphology and densification during sintering, which can lead to a more pronounced grain growth effect due to enhanced diffusion pathways created by these defects [15].

Composite Materials: SEAJCCM2024   Materials Research Forum LLC
Materials Research Proceedings 56 (2025) 117-123   https://doi.org/10.21741/9781644903636-13

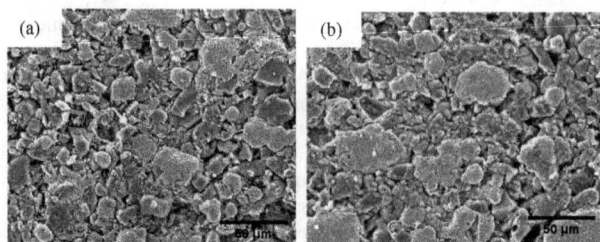

Fig. 2. Microstructure of sintered (a) CuAlSi alloy without ball mill and (b) CuAlSi alloy after 3 hours of ball mill.

EDX of CuAl and CuAlSi Alloys

The EDX results provide insights into the compositional uniformity of CuAl and CuAlSi alloys. Fig. 3 illustrates the elemental distributions for 3(a) unprocessed CuAl alloy, 3(b) ball-milled CuAl alloy, 3(c) unprocessed CuAlSi alloy, and 3(d) ball-milled CuAlSi alloy. In Fig. 3(a), Cu is represented in red and Al in green. The unprocessed CuAl alloy shows a homogenous distribution of elements with fine grains as mentioned before, there is no stored energy to aid the merging of smaller grains into larger ones. However, Fig. 3(b), which represents the ball-milled and sintered CuAl alloy, reveals significant changes in microstructure. After mechanical alloying and thermal treatment, red regions now represent Al, and green regions represent Cu. The thermal energy provided during sintering enhances atomic mobility, allowing atoms to move more freely. The combination of thermal energy and the pre-existing defect from mechanical alloying promotes rapid migration of grain boundaries. This migration is crucial for grain coalescence, where smaller grains merge into larger ones [16].

Fig. 3(c) shows the distribution in the unprocessed CuAlSi alloy, where Cu, Si, and Al are represented by green, blue, and red, respectively. The elements exhibit a well-distributed composition, attributed to their equimolar ratios. The uniformity of the distribution highlights the effectiveness of the initial powder mixing in achieving compositional balance. In contrast, Figure 3(d), which shows the ball-milled CuAlSi alloy, the same colour scheme used for Si, Cu, and Al reveals changes after sintering. Furthermore, the coarsening observed in the CuAlSi alloy reflects grain coalescence, where the energy from mechanical alloying and thermal exposure enhances grain boundary mobility, allowing grains to merge and grow, thereby influencing the mechanical properties of the alloy [17].

The presence of Si led to a reduction in Cu after the ball was milled and sintered. This is due to Si reacting with oxygen to form silicon oxides which lead to a localized depletion of oxygen around Cu. In a study of Cu-rich precipitates in silicon, it was indicated that Si tends to form stable bonds with oxygen, which influences the overall oxidation behavior of Cu in its vicinity. It was noted that while Cu tends to form silicides, it can also be oxidized under certain conditions due to the presence of oxygen and the thermodynamic stability of copper oxides compared to silicides [18]. Other research on the high-temperature oxidation behavior of iron-based alloys containing Cu and Si indicates that Si affects the oxidation of Cu-rich phases [19].

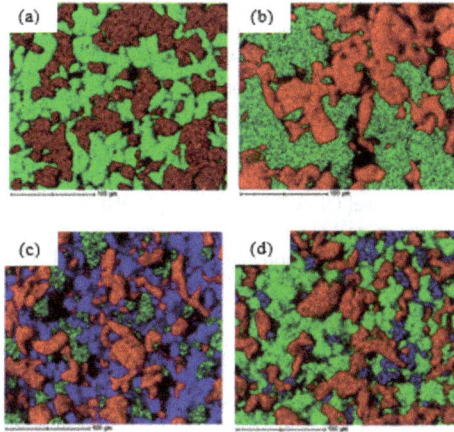

Fig. 3. Elemental distribution of (a) as-is CuAl alloy, (b) 3 hours ball milled CuAl alloy, (c) as-is CuAlSi alloy, and (d) 3 hours ball milled CuAlSi alloy.

Conclusion

In conclusion, this study investigates the effect of Si addition and the effect of mechanical alloying on CuAl and CuAlSi alloys using the powder metallurgy technique. The addition of Si and ball milling promotes grain coalescence. This effect leads to coarser and more elongated grains compared to finer grains in unprocessed CuAl powders. The main reason is that this process introduces defects and stored energy in CuAl and CuAlSi alloys, enhancing atomic mobility during sintering. This promotes grain coalescence where smaller grains merge into larger grains. This also affects grain compactness and refining elemental distribution. EDX results show that the presence of Si in CuAlSi alloys reduces the Cu content after sintering due to the formation of silicon oxides.

Acknowledgments

The authors would like to acknowledge the support from the Fundamental Research Grant Scheme under grant number FRGS/1/2022/STG05/UNIMAP/02/3 from the Ministry of Education Malaysia, Center of Excellence Geopolymer & Green Technology (CEGeoGTech), Faculty of Chemical Engineering and Technology, Universiti Malaysia Perlis (UniMAP) for their partial support.

References

[1] L.G. Sun, G. Wu, Q. Wang, J. Lu, Nanostructural metallic materials: Structures and mechanical properties, Materials Today 38 (2020) 114-135. https://doi.org/10.1016/j.mattod.2020.04.005

[2] Z. Ma, K. Zhang, Z. Ren, D.Z. Zhang, G. Tao, H. Xu, Selective laser melting of Cu–Cr–Zr copper alloy: Parameter optimization, microstructure and mechanical properties, Journal of Alloys and Compounds 828 (2020) 154350. https://doi.org/10.1016/j.jallcom.2020.154350

[3] D. Varshney, K. Kumar, Application and use of different aluminium alloys with respect to workability, strength and welding parameter optimization, Ain Shams Engineering Journal 12 (2021) 1143-1152. https://doi.org/10.1016/j.asej.2020.05.013

Composite Materials: SEAJCCM2024   Materials Research Forum LLC
Materials Research Proceedings 56 (2025) 117-123   https://doi.org/10.21741/9781644903636-13

[4] A. Fadillah, P. Puspitasari, A.A. Permanasari, S. Sukarni, R. Nurmalasari, A. Wahyudiono, Analysis of tensile strength, hardness, and fracture surface of aluminum silicone (Al-Si) with various copper (Cu) percentages using stir casting method, AIP Conference Proceedings 2687 (2023) 1. https://doi.org/10.1063/5.0120994

[5] D.Y. Ying, D.L. Zhang, Solid-state reactions between Cu and Al during mechanical alloying and heat treatment, Journal of alloys and compounds 311 (2000) 275-282. https://doi.org/10.1016/S0925-8388(00)01094-X

[6] B. Bukola Joseph, B. Michael Oluwatosin, B. Joseph Olatunde, A. Kenneth Kanayo, Corrosion characteristics of as-cast aluminium bronze alloy in selected aggressive media, Journal of Minerals and Materials Characterization and Engineering 1 (2013) 245-249. https://doi.org/10.4236/jmmce.2013.15038

[7] J. Tao, K. Yang, H. Xiong, X. Wu, X. Zhu, C. Wen, The defect structures and mechanical properties of Cu and Cu–Al alloys processed by split Hopkinson pressure bar, Materials Science and Engineering: A 580 (2013) 406-409. https://doi.org/10.1016/j.msea.2013.05.067

[8] V. Rajkovic, D. Bozic, J. Stasic, H. Wang, M.T. Jovanovic, Processing, characterization and properties of copper-based composites strengthened by low amount of alumina particles, Powder Technology 268 (2014) 392-400. https://doi.org/10.1016/j.powtec.2014.08.051

[9] M. Gazizov, C.D. Marioara, J. Friis, S. Wenner, R. Holmestad, R. Kaibyshev, Precipitation behavior in an Al–Cu–Mg–Si alloy during ageing, Materials Science and Engineering: A 767 (2019) 138369. https://doi.org/10.1016/j.msea.2019.138369

[10] X.M. Pan, C. Lin, H.D. Brody, J.E. Morral, An assessment of thermodynamic data for the liquid phase in the Al-rich corner of the Al-Cu-Si system and its application to the solidification of a 319 alloy, Journal of phase equilibria and diffusion 26 (2005) 225-233. https://doi.org/10.1007/s11669-005-0109-1

[11] M. Weigl, F. Albert, M. Schmidt, Enhancing the ductility of laser-welded copper-aluminum connections by using adapted filler materials, Physics Procedia 12 (2011) 332-338. https://doi.org/10.1016/j.phpro.2011.03.141

[12] J.B. Fogagnolo, E.M. Ruiz-Navas, M.H. Robert, J.M. Torralba, The effects of mechanical alloying on the compressibility of aluminium matrix composite powder, Materials Science and Engineering: A 355 (2003) 50-55. https://doi.org/10.1016/S0921-5093(03)00057-1

[13] W. Luo, Y. Guo, Z. Xue, X. Han, Q. Kong, M. Mu, G. Zhang, W. Mao, Y. Ren, Microstructure and mechanical properties of the Mg–Gd–Zn alloy prepared by sintering of rapidly-solidified ribbons, Scientific Reports 12 (2022) 11003. https://doi.org/10.1038/s41598-022-14753-2

[14] B.N. Sharath, R.R. Rao, K.S. Madhu, S. Pradeep, Sintering of mechanically alloyed powders, Advancements in Powder Metallurgy: Processing, Applications, and Properties (2024) 188-222. https://doi.org/10.4018/978-1-6684-9385-4.ch008

[15] P. Kulecki, E. Lichańska, The effect of powder ball milling on the microstructure and mechanical properties of sintered Fe-Cr-Mo-Mn-(Cu) steel, Powder Metallurgy Progress 17 (2017) 82-92. https://doi.org/10.1515/pmp-2017-0009

[16] Q. Guo, H. Hou, K. Wang, M. Li, P.K. Liaw, Y. Zhao, Coalescence of Al0. 3CoCrFeNi polycrystalline high-entropy alloy in hot-pressed sintering: a molecular dynamics and phase-field study, Npj Computational Materials 9 (2023) 185. https://doi.org/10.1038/s41524-023-01139-9

Composite Materials: SEAJCCM2024
Materials Research Proceedings 56 (2025) 117-123

Materials Research Forum LLC
https://doi.org/10.21741/9781644903636-13

[17] S.D. Gaikwad, P. Ajay, V.V. Dabhade, S.N. Murty, S. Manwatkar, U. Prakash, Mechanical properties and microstructural analysis of ultra-fine grained Ni-based ODS alloy processed by powder forging, Journal of Alloys and Compounds 970 (2024) 172614. https://doi.org/10.1016/j.jallcom.2023.172614

[18] T. Buonassisi, M.A. Marcus, A.A. Istratov, M. Heuer, T.F. Ciszek, B. Lai, Z. Cai, E.R. Weber, Analysis of copper-rich precipitates in silicon: Chemical state, gettering, and impact on multicrystalline silicon solar cell material, Journal of Applied Physics 97 (2005) 063503. https://doi.org/10.1063/1.1827913

[19] B.A. Webler, S. Sridhar, The effect of silicon on the high temperature oxidation behavior of low-carbon steels containing the residual elements copper and nickel, ISIJ international 47 (2007) 1245-1254. https://doi.org/10.2355/isijinternational.47.1245

Keyword Index

About the Editors

Norhayani Othman is a senior lecturer at Faculty of Chemical and Energy Engineering, Universiti Teknologi Malaysia (UTM). She also a researcher in the Enhanced Polymer Research Group (EnPro), an associate member of Centre for Advanced Composite Materials (CACM) and a member of Material and Manufacturing Research Alliance (MMRA). She has a first degree and a master degree in Polymer Engineering from UTM and received her PhD in Chemical and Biological Engineering from The University of British Columbia, Canada in the year of 2012. Her master degree thesis is on characterisation of toughened polyamide 6/polypropylene nanocomposites. During PhD she studied the melt and processing behaviour of polylactic acid and their enantiomeric copolymers and blends. Her current research interest includes plastic waste management, crystallisation and degradation behaviour of recycled plastics and biopolymers, melt rheology and processing. She has published more than 50 research papers and conference proceedings on related areas.

Lin Feng Ng is currently a postdoctoral researcher in the Faculty of Mechanical Engineering, Universiti Teknologi Malaysia (UTM), Malaysia. He completed his PhD degree in 2021 at Universiti Teknikal Malaysia Melaka, Malaysia. He received the PhD Best Thesis Award from the Universiti Teknikal Malaysia Melaka in 2022. During his postgraduate study back in 2015, he earned a silver award at the UTeM Research and Innovation Expo. To date, he has published 45 journal articles, 2 conference papers and 18 book chapters indexed by Scopus/ISI. His publications have been cited more than 800 times, bringing the h-index to 19 and the i-10 index to 32. In 2024, he was awarded the IOP Outstanding Reviewer award for reviewing the Engineering Research Express. He also received the IOP Trusted Reviewer Status in 2023. He is currently holding a post as committee member of several international conferences. He is also serving as a reviewer for several high-impact and reputable journals. In all, he has reviewed more than 300 research articles that have been submitted to those high-impact journals. His research interests include Fibre-metal laminates, Polymer composites, Hybrid composites, Synthetic/Natural fibres, Mechanical characterisation, Fatigue life assessment and Energy absorption.

Pui San Khoo is currently a postdoctoral researcher at the Centre for Advanced Composite Materials, Universiti Teknologi Malaysia. She was awarded a scholarship by the Public Service Department (JPA) in 2012 to pursue a Bachelor's degree in Wood Science and Technology at Universiti Putra Malaysia (UPM), which she successfully completed in 2015. Following her undergraduate studies, she received the MyBrain15 sponsorship from the Malaysian government to pursue a Ph.D. in Biocomposite Technology and Design at the Institute of Tropical Forestry and Forest Products (INTROP), UPM. Her research interests encompass wood machining, biocomposites, material science, and polymer composites. She has authored or co-authored a total of 52 publications, including 40 journal articles indexed in JCR/Scopus, one non-indexed journal, six book chapters, and five Scopus-indexed conference proceedings, primarily focusing on biocomposites and biopolymers.

R.A. Ilyas is a senior lecturer in the Faculty of Chemical and Energy Engineering, Universiti Teknologi Malaysia, Malaysia. He is also a Fellow of International Association of Advanced Materials (IAAM), Sweden, Fellow of International Society for Development and Sustainability (ISDS), Japan, a member of Royal Society of Chemistry, UK and Institute of Chemical Engineers (IChemE), UK, Chair of Science Outreach for Young Scientists Network - Academy of Sciences Malaysia (YSN-ASM) 2023. He received his Diploma in Forestry, Bachelor's

b

Degree (BSc) in Chemical Engineering, and Ph.D. degree in the field of Biocomposite Technology & Design at Universiti Putra Malaysia, Malaysia. R.A. Ilyas was the recipient of the MVP Doctor of Philosophy Gold Medal Award UPM 2019, for Best Ph.D. Thesis and Top Student Award, INTROP, UPM. He was awarded with National Young Scientist Award 2024 (Engineering) by Ministry of Science, Technology and Innovation (MOSTI), Malaysia, Outstanding Reviewer by Carbohydrate Polymers, Elsevier United Kingdom, Top Cited Article 2020-2021 Journal Polymer Composite, Wiley, 2022, and Best Paper Award at various International Conferences. R.A. Ilyas also was listed and awarded among World's Top 2% Scientist (Career-Long Achievement) Year 2022 and 2023, World's Top 2% Scientist (Subject-Wise) Citation Impact during the Single Calendar Year 2019-2023 by Stanford University, US, PERINTIS Publication Award 2021 and 2022 by Persatuan Saintis Muslim Malaysia, Emerging Scholar Award by Automotive and Autonomous Systems 2021, Belgium, Young Scientists Network - Academy of Sciences Malaysia (YSN-ASM) 2021, UTM Young Research Award 2021, UTM Publication Award 2021&2023, and UTM Highly Cited Researcher Award 2021. In 2021, he won Gold Award and Special Award (Kreso Glavac (The Republic of Croatia) at the Malaysia Technology Expo (MTE2022), Gold Award dan Special Award at International Borneo Innovation, Exhibition & Competition 2022 (IBIEC2022), and, a Gold Award at New Academia Learning Innovation (NALI2022). He was awarded with Best Scientific Book Award from COMSTECH, Organization of Islamic Cooperation (OIC), Pakistan and ModTech, Romania. His main research interests are (1) Polymer Engineering (Biodegradable Polymers, Biopolymers, Polymer composites, Polymer-gels) and (2) Material Engineering (Natural fiber reinforced polymer composites, Biocomposites, Cellulose materials, Nano-composites). To date he has authored or co-authored more than 500 publications on green materials related subjects.

Mohd Yazid Yahya is a Professor at the Faculty of Mechanical Engineering, Universiti Teknologi Malaysia, and holds a PhD in Composite Structure from the University of Liverpool. He is the Director of the Centre for Advanced Composite Materials (CACM), a specialist in Composite Structure, Pressure Vessels, Impact, and Blast loading.

www.ingramcontent.com/pod-product-compliance
Lightning Source LLC
Chambersburg PA
CBHW071706210326
41597CB00017B/2365